Cyrus Mason Tracy

Studies of the Essex Flora

An Enumeration of all Plants Found Growing Naturally within the Limits of

Lynn, Mass.

Cyrus Mason Tracy

Studies of the Essex Flora
An Enumeration of all Plants Found Growing Naturally within the Limits of Lynn, Mass.

ISBN/EAN: 9783337270070

Printed in Europe, USA, Canada, Australia, Japan

Cover: Foto ©berggeist007 / pixelio.de

More available books at **www.hansebooks.com**

STUDIES

OF THE

ESSEX FLORA:

An Enumeration of all Plants found Growing
Naturally within the Limits of Lynn, Mass.,
and Towns adjoining; with Notes
as to Localities and Habits.

BY

CYRUS MASON TRACY.

———

LYNN, MASS.
THE NICHOLS PRESS — THOS. P. NICHOLS.
1892.

Dedication.

In sweet memory of

LONG WALKS THROUGH FERN-LINED PATHS,
OF FLOWER-LORE TAUGHT IN COOL SHADOWS,
AND IN
RICH ANTICIPATION OF FUTURE STUDY IN FAIRER FIELDS,
THIS SECOND EDITION OF THE AUTHOR'S STUDIES
IS SENT FORTH BY

HIS CHILDREN.

PREFACE.

DURING all the intervening years between the first publication of these studies and the present edition, it was the strict habit of the author to note every new discovery, every change in locality, preparatory to a second issue. These notes were written in a close hand, crowded into the uncut margin of an old copy. About two weeks previous to Mr. TRACY'S death he decided on immediate publication. Feeling incompetent to undertake the classification of the Grasses and Sedges, he opened correspondence with several, hoping to find among them one to whom he could trust this part of the work; finding no one giving satisfaction he began the preparation of the manuscript, and had written all the headings, carrying them through the *Endogens;* how much further he had decided to carry the work alone we are ignorant; we have chosen, however, to leave it where he left it rather than that another should add work, perhaps of a far different character.

This explanation will make plain our reason in bringing out first a book requiring re-writing rather than one of the many already prepared. It was first in his thought and plan, therefore first in our execution.

S. E. T.

Lynn, March 19, 1892.

ERRATA.

Page 25 *for* Cardimine, *read* "Cardamine."
" 32 *for* Celastrus scandeus, *read* "scandens."
" 38 *for* Amelanchiar, *read* "Amelanchier."
" 39 *for* Ludwigea, *read* "Ludwigia."
" 41 *for* Saxafrids, Saxafrage, *read* "Saxifrids, Saxifrage."
" 43 *for* Canium, *read* "Comium."
" 44 *for* Turritus, *read* "Turritis."
" 48 *for* Rosens, *read* "Roseus."
" 57 *for* Galutheria, *read* "Gaultheria;" Azalea, "Azalia."
" 66 *for* Echinospernum, *read* "Echinospermum."
" 67 *for* Solonum, *read* "Solanum."
" 70 *for* Ameranthus, *read* "Amaranthus."
" 74 *for* Q. bicolar, *read* "bicolor."
" 80 *for* Alisma, *read* "Alisima."

Page 73 *after* E. polygonifolia, *insert*

Spotted Spurge, **E. maculata.** L.
Milk Purslane.
 Common. A flourishing weed in most gardens. Perhaps the only native species in this vicinity.

Page 79 *after* P. natans, *insert*

Short-spiked **P. hybridus.** Michx.
Pondweed.
 Occasional. Still, clear waters.

PREFACE.

DURING all the intervening years between the first publication of these studies and the present edition, it was the strict habit of the author to note every new discovery, every change in locality, preparatory to a second issue. These notes were written in a close hand, crowded into the uncut margin of an old copy. About two weeks previous to Mr. TRACY's death he decided on immediate publication. Feeling incompetent to undertake the classification of the Grasses and Sedges, he opened correspondence with several, hoping to find among them one to whom he could trust this part of the work; finding no one giving satisfaction he began the preparation of the manuscript, and had written all the headings, carrying them through the *Endogens;* how much further he had decided to carry the work alone we are ignorant; we have chosen, however, to leave it where he left it rather than that another should add work, perhaps of a far different character.

This explanation will make plain our reason in bringing out first a book requiring re-writing rather than one of the many already prepared. It was first in his thought and plan, therefore first in our execution.

S. E. T.

Lynn, March 19, 1892.

INTRODUCTION.

No one who has followed rare old Gilbert White through his Natural History of Selborne, who has read the History of the Bass Rock, or the later volumes of Thoreau,

"Tasting of Flora and the country green,"

will ask for any labored reason why this little work has been undertaken. Those authors have abundantly shown the pleasant and entertaining nature of local research and description; and we only need look to the enduring reputation of the *Florula Bostoniensis*, as a practical work, to complete the argument, and assure us of the sufficiency of a limited territory to furnish material for profitable thought for a long period of time.

To Essex County, Mass., may be applied the full force of the remark of Bigelow, that "the Flora of any considerable section of our territory may furnish full occupation for years." It has been the scene of the pioneer labors of Cutler and the thorough operations of Oakes; it has given employment by turns to the scrutinizing eyes of Osgood, Nichols and Russell, as well as many others; yet not a few of the localities of this small district seem to remain comparatively unexplored.

The southwestern corner of the County, occupied by the townships of Lynn, Saugus, Lynnfield, Swampscott and Nahant, the present separated fragments of ancient Lynn, would appear to be one of the more neglected of these portions. Those who love pleasant and finely-toned scenery have often found much satisfaction in this vicinity, and the culler of choice old histories and romantic legends has long esteemed it a productive field; but the practical botanist seems, for the most part, to have preferred to explore the vegetation of Cape Ann, to turn his pilgrimage to the White Mountains, or drive his hunt through the woods of Middlesex or the meadows of Plymouth, rather than believe that a district so near the metropolis might contain some things worth looking for. There would, perhaps, be no propriety in saying that these other sections are not, any of them, very much superior in natural riches to this; but I strongly suspect that the study of the flora, which is the subject of this work, may reveal phenomena and peculiarities of vegetation, calculated to interest even an adept.

A rather striking diversity, both in geological and botanical productions, appears in the territory under notice.

From the western part of Swampscott an extensive formation of porphyry begins, and sweeping its northern limit along a gently-curving line, follows the valley occupied by the well-known "Lakes of Lynn" as far as the Sluice Pond, thence traverses the township of Lynn through its center, very nearly, and by a direction not far from east and west; then, passing the valley of Saugus River, in the neighborhood of Pranker's Factories, trends gradually to the southwest and is lost in the hills of Malden and Medford. South of this line there is hardly anything but porphyry to be found in place; to the north there is next to none of it, but the region has all the usual characteristics of one which rests almost wholly on granite.

As any one versed in the subject of natural scenery might expect, the aspects of these two divisions of country are widely different. In the southerly or seaboard portion, the bold eminences of High Rock, Sadler's Rock, Lover's Leap, Forest Rock, and others, well exemplify the prominent traits of the porphyry; hard, stern and precipitous on the southern side in almost every case, looking with inflexible front toward the sea, as if they were the stout old knights that in ancient time had driven back the onset of its marauding waters; and on the north as uniformly smoothed and rounded, shelving back with a gentle slope, and sinking in the yellow soil of the hills.

The northern section, so far as it is embraced within the boundaries of Lynn, forms one spacious common forest, known as the Ox Pasture; a district where Nature seems to have dallied long and wondrously with the giants of the age of granite.

Long, deep and solid ledges furnish block after block to reward the patience of the quarrymen; and here and there their gaping pits in the hillsides afford a partial sight into one of the many caskets in which New England stores her jewels. But older and sturdier quarriers have wrought here,—the stupendous crush, and jar, and rend of the drift period have seemingly tossed the fragments, of hundreds of tons weight, like footballs, leaving them in some instances perched on the brink of precipices, in what would appear the most unstable attitudes, or again, scattering them over the hill-slopes, small and great together,

> "Thick as autumnal leaves that strew the brooks,
> In Vallambrosa."

The chemist avers that to cultivate any crop successfully in the field, a studied adaption of the soil to the particular plant in view must ever be made. The converse of this rule would indicate that special characters existing naturally in a soil should give corresponding differences in the kind and style of vegetation which it produces. If I wished for an illustration of this idea, I could hardly find a better case than appears in the two formations under notice.

In passing through Lynn Woods it is not difficult to detect, even with small experience, the exact line of junction of the granite and porphyry, within a few rods, by the style of vegetation alone. A few examples will make this more definite.

On the rocky pasture hills that immediately overlook the city of Lynn, the Barberry starts in unrestrained abundance, the Privet adorns whole acres in early summer with its little clusters of snowy flowers, and the Pitch Pine and Red Cedar assert their right to the land with the vigor of feudal barons. When we pass northward over this natural mark, the Privet disappears almost entirely, the Barberry becomes the exception instead of the rule, the Cedars are scarce and the multitudes of Pitch Pine are only represented by a few stragglers. To replace them, however, the Beech, of which only two specimens grow on the porphyry to my knowledge, and these I suppose to be artificially located, starts up at once almost on the very boundary, and stretches away from thence in vigorous condition towards the woods of Lynnfield. The Chestnut, that joy of country lads and squirrels, ventures down into the north of Saugus in commendable strength, but cannot cross the enchanted line without the help of man, and in cultivation grows slowly and timidly, as if it were ill at ease. More remarkable than either, the Black Larch or Hacmatac. which, I venture to say, is unknown as a native south of the granitic section, is found growing and thriving within fifty rods of its margin. The Blue Vervain, the Water Avens, and the Mountain Mint, look doubtfully in from the east over the channel of Stacy's Brook, but effect no further progress, and on the west the Knapweed, which revels by the wayside on the Chelsea hills, makes no attempt to establish itself on ground from which it is so singularly debarred.

If my enumeration of the plants of Lynn and vicinity serves no other purpose than to develop a phenomenon so interesting as the above, it will still be far from a vain undertaking. But this is by no means all the motive for the enterprise.

> " There is a pleasure in the pathless woods;
> There is a rapture on the lonely shore."

To him who loves Nature for her own sake she ever dispenses rewards more precious than gold. Botanical pursuits, though harmonizing well with activity and energy of temperament. are yet potent to soothe and tranquilize the fretted spirit; they have all the quieting power which Mudie ascribes to moonlight. I do not wonder that the lamented Oakes, disturbed by the discordance of the jarring interests and conflicting elements of the law, should have taken sanctuary where

> " The groves were God's first temples."

Moore, when he wrote of the bowers

> " Where Pleasure lies, carelessly smiling at Fame,"

drew but a very dim outline of that luxurious satisfaction which one feels, when. lounging down with his favorite specimens at noon, on the grassy banks of some merry little brook, with the thousand vertical shadows gamboling among the ferns around him, he gives himself up to the full inspiration of the place, and, hardly capable of a craving, is but too well satisfied to watch the water-spider that dangles in his flimsy web from rush to rush, or the greedy emmets that hunt among the grass-blades for the crumbs of his slowly-vanishing biscuit. Or, changing the stream for the brink of some pond between the hills, his contemplation alternates from the swift ripples, that break and mutate like the figures in a kaleidoscope, to the upland banks of pleasing contour, spread with dense verdure and counterwrought in the water beneath, or to the soft light and shade that blend through the rounded masses of clumps of oak and hickory, or give additional life to long lines of " willows by the watercourses."

But botany in practice is very far from being all indolence. There are long jaunts to be taken, where the horse and carriage would fare badly and work but ill. There are hills to be climbed and tall rocks to be scaled for the prize that hangs its attractions from crevices high above reach. There are swamps to be penetrated, where the feet must risk a wetting and mock at mire; and thickets to be searched with as much carefulness as though upon a legal warrant, though the proper results of the effort may appear when the garments, like the galligaskins in the Splendid Shilling,

"A horrid chasm disclose."

Yet how little of all this does the mind regard, at such moments as I myself have seen, when the end *had crowned* the work, and the object of my search was before me, humble and unpretending sometimes, yet often glowing in all the excellence of its floral dignity. I well remember a hot and weary day when, late in the afternoon, I was coursing with others over the rugged hills and crags of Malden. Tired and thirsty, I was inwardly complaining of the toilsome and profitless route, when, leaping down from a rough pole-fence, I stood face to face with the most magnificent Oak-leaved Gerardia I ever saw. Had the wealth of its yellow bells been coined to very gold in my hand, I could have felt no higher satisfaction than I had in seeing its four-foot stem, crowded with brilliant flowers, swaying to and fro in the warm westerly wind, the magic wand to charm away, for the time, every thought of fatigue. In a certain summer I had a kind of botanical vow, which I kept long inviolate, to let no day pass without the determination of at least one new species. I was fresh in the study then, and such an idea was nowise absurd. But one day had waned until the sun had actually gone down on my errantry, which threatened to become night-errantry, sure enough. A boggy meadow, often visited before, seemed the only available spot, and to it I turned with the resolution of a forlorn hope. Fifteen minutes later, had my feet responded to my feelings, I should have been dancing among the hassocks for the discovery of the charming Cymbidium, which I had not seen before since I gathered its blushing beauties when a boy, in the meadows of Connecticut.

When I first angled the Purple Bladderwort from its oozy couch, and told my comrades that the sight of it was worth a dollar, I was only laughed at for my enthusiasm. When I found a Corydalis magnificently growing and blooming on the slope of Dungeon Rock, and detaching the whole bunch, earth and all, carried it home in my arms, that the beautiful specimen might not be injured, no one could see any good reason for my lugging so much dirt for a few flowers. I was hardly sensitive enough on the point to commence an argument: I had gained a treasure that flowered all summer for me where I set it; and long before its seeding-time, I had forgotten the wet feet I got in Pine Hill Swamp when bringing it home.

In introducing the list of plants which follows, I would be glad to incorporate into it some quality to give it the zest for others which the originals have had for me. Vitality of expression and the hue of health do not appear in the most perfect statuary; and one may read a simple enumeration of vegetable forms, and feel nothing of the fresh winds that bend the tree-tops — see nothing of the scenic effect of spreading verdure, piled-up masses of foliage, or hill beyond hill, stretching away through the softening gradations of distance.

This little district has no Merrimac or Hudson sweeping round its borders,—no Kearsarge or Wachusett standing for its protector; yet much of the higher and purer delights of open-air research, may always be enjoyed in its exploration. There are peculiarities in every spot, and this is one where a strong feature is seen in the close connection of seaside and wood scenery; closer, perhaps, than in most other places. Suppose then, reader, we walk down to the shore. You are reading this, probably, because you like plants—Nature in vegetative attire—and like them well enough, I trust, to go where they grow by means of your own powers, not waiting to call for a horse and chaise. To Long Beach, then, that rope of sand, but wonderfully durable, by which, as the clue of Theseus, the young town of Nahant feels its way back now and then to the lap of old Lynn, its mother for two centuries. Here, where the grand, long swell comes rolling heavily in from the middle deeps of the Atlantic and breaks at our feet, like the utterance of great thoughts in human language as near as may be, and yet too often misinterpreted—here we can stand among rocks thickly fringed with dripping seaweeds, or crusted with barnacles and studded with creeping shell-fish, and look away to the iron headlands of East Point and the Spouting Horn, where the Beach Pea straggles among the pebbles and the Pimpernel hangs in the chinks of the rock

     "Half way down"
   Like "one that gathers samphire."

  Threading our way eastward along the craggy shore, we may thorn our fingers at pleasure with the prickly Saltwort, or treat our palates to an equal pungency in the radish-like flavor of the Sea-Rocket; and thus we round the jagged projections, and pass the many springs of pure fresh water that bubble up so generously within the tide-range, till a bright stream comes dancing down to meet us, and we give a hand in greeting to Stacy's Brook and the thrifty village of Swampscott at one and the same time.
  Such as have not seen Stacy's Brook must not imagine that it differs, materially, from other New England rivulets. Like them it has its depths and its shallows, its sudden angles and smooth reaches, its meadow banks overhung with alders and willows, where the mole and muskrat love to mine, and its impulsive gushings under little road-bridges, where the water goes sparkling over red and white pebbles, and the cows come sedate-

ly down to drink at noon. The stream runs, as you see, reader, through open meadows mostly, till we are near a mile from the shore, when we find it pressing through a culvert in the heavy embankment of the Eastern Railroad, and passing this we are at once welcome to the hospitalities of Linnean Grove.*

The pleasant little copse, whose name hints so strongly at its character, has been so christened by the lovers of botany in its neighborhood, on account of the variety and peculiarity of its productions. We perceive that it consists entirely of hardwooded trees, partly on upland and partly on low ground, with the brook traversing the western margin. Many plants are here that we shall find nowhere else in the district, while for many others it is the best and most accessible locality. I do not know the owner of Linnean Grove. Whoever he is, let him have the praise of preserving to us and others one delightful relic of Nature among the devastations of improvement. May his shadow never be less, nor that of his grove.

Let us go on tracing up the course of the brook. We cross Essex Street and find it turning a mill, the only time it is pressed into such ignoble service. Above this, it comes lazily down through the mossy hollows of Fresh Marsh, where the Fringed Gentian and Grass-of-Parnassus love its copious moisture, and the Arethusas and Cymbidium go, as it were, strolling about among the sedgy grass even to the base of Rocks Pasture, "ye Woodende Rocks" of the old records, that rise so sturdily on the west. Before us is Gravesend, the Hamlet of the Lakes, where the spirit of rural beauty and quietness lingers yet, though expelled from almost all the rest of the township. Let us contemplate it while we may, for Young America rides an iron steed, and when we come again we may find the mantle of Retirement torn and trampled under his impetuous hoofs. The brook betrays its origin here in the Floating-Bridge Pond, long reputed bottomless and no doubt of extreme depth in the center, but certainly quite shallow toward the southern end. Here we may pass out on the low bridge that, bare of romantic arch or ivied buttress, lies stretched out on the water as though it were

---

* Linnean Grove, together with several other localities mentioned in the first edition of these studies, has been sacrificed to the cheap house-builder, and can no longer be recognized by the botanist. — ED.

the torpid whale of Sinbad, and look down among the leaves of the water lilies, or see the Button-bush and Sweet-gale wading out into deep water side by side with the pearl hunters, though not like those, pulling up the unoffending shell-fish to gratify a feverish desire.

By a short walk northwesterly, we shall reach the Flax Pond, or, as it will be found marked on the map, Wenuchus Lake. This is by far the finest sheet of water in Lynn, and if we except Spring Pond in the edge of Salem, there is nothing superior nearer than Lynnfield. Receiving the stream from the other ponds on the north, it sends out the perennial current of Strawberry Brook towards the southwest, first through the Bowler Swamp, and thence along a charming valley, quite to the western line of the city.

On the eastern side of the Flax Pond rises a picturesque little eminence known as Silsbee's Hill. From the top of this we can enjoy as pleasant a view towards the west as we can well desire; or, turning northeastward, we may at our leisure study the indications of the granite and porphyry, as visible, on the right and left, in the style of surface and vegetation.

We will pass by Flax Pond on the eastern side and follow the valley northerly to the Sluice Pond. Here is another mill-site; and the little village round it has an air of manufactures. We still continue northerly, for the pond is long and narrow, making its quiet bed all the year in this little hollow, between Indian Hill on one side and the east flank of the Blood's Swamp Hills on the other. When we come to the northern point we shall find the water oozing in on all sides from the gravelly banks, and in spring a gay rivulet rattling along from Spruce or Cedar Pond, still farther north, and falling in to augment the waters of "ye olde Sluice."

At this point a new prospect opens. We have reached the limit of the streams that flow southerly, and after this shall find them traveling, like the star of empire, westward. The ridge which the Lynnfield road here occupies is identical with that of the Blood's Swamp Hills, this being its lowest point. Do you observe, reader, the gorge-like valley that opens towards the west? Do you remark the heavy shadows of the pine woods on the southern declivity, contrasting so agreeably with the lighter spray and more undulating foliage on the northern side? Notice, also, how the diverse outlines of the strong ledges and thickly-wooded hills are seen merging into one another, and as they retire more and more, gray succeeds to green, bluish to gray, and this again deepens to a full azure. This is the most remarkable depression in Lynn: the valley of Blood's Swamp.

Entering by a secluded but pleasant wood-road, we will follow its somewhat devious course up the hill to the left. If you are not in haste, we can regale ourselves with checkerberry leaves, or stay to gather the box-berries that are strewn about as if the drops of dew had been congealed into coral. When we come here in the spring the Blue Hepatica will now and then look up from the damp leaves and give us a quiet welcome; and if we repeat the visit somewhat later, the Linnea will scatter a pure incense from the twin bells of its woodland censer, and be ready to receive us in its cool arbors, whence it never ventures into the open sunshine. Or we may here and there chance upon a Desmodium; or a bushy Gerardia, as free in its beauty as when it first glowed on the botanical genius of Oakes, that wakened to life before it, as its own buds before the vernal sun. This open spot through which we are passing, will by and by be spangled over with the Yellow Star-of-Bethlehem, and a little later the lithe Golden-Rods will nod and beckon to each other among the larger trees and between the scrub oaks and sassafras bushes. Perhaps they will whisper: I cannot tell.

We are now on the highest of the Blood's Swamp Hills; almost, if not quite, the highest land in the township. There is an unbroken horizon, taking in the long ocean sweep from Boston Harbor to Cape Ann, and then the highlands of Central Essex, including those of Topsfield, Boxford and Andover, with Reading almost at our feet, and the wood-covered eminences rolling beyond, like billows in an ever-heaving ocean, till the prospect fades away among the alpine highlands of southern New Hampshire, with Monadnock and its co-ordinate peaks for the termini. When the weather is thick or the northern sky hazy, nothing of these is to be seen; but in a clear air, such as we have to-day, you observe how lightly and cloudlike their immense forms seem to float, or rather hover, on the horizon. I once stood on this rock, with a friend as enthusiastic as myself, just when the sun was setting, and the long level beams of yellow light were shooting over the tree-tops in the swamp below us like slender arrows. The sky, for the most part, was free from vapors; but low down in the far north, beyond all else, there hung a dense bank of clouds, dun-colored and sluggish, against which the deep blue of those granite hills lay clear and sharply drawn, more clear, indeed, than the outline of hills five miles away.

There is something amazingly grand in the fair view of a mountain, even when it is far off and comparatively small. My friend was more familiar with alpine scenery than I; he had travelled and dwelt in its midst; yet so forcible and purely elemental was the scene before us, that he, no less than I, stood like

"A pensive pilgrim at the foot
Of the crowned Alleghany, when he wrapt
His purple mantle gorgeously around,
And took the homage of the princely hills
That stood before him, as they bowed them down
Each in his order of nobility."

Below us, on the north, the gorge of the swamp sinks down like the entrance to some mighty cavern; and tracing it westerly, you can detect its egress, where the hills are exchanged for the Saugus meadows. Beyond are the high ridges of woods of eastern Middlesex, and Wachusett rising above the whole. Southward from these, the eye ranges over the gravelly hills of Chelsea and Somerville, with the swarming hive of the capital, veiled and panting in the smoke of furnaces and railroad stations.

Our stay in this free and uncontaminated air has probably refreshed you, reader, so that you are no wearier than I. A short walk southwesterly brings us to Dungeon Rock; the place of much that is beautiful, much that is fanciful, much that is perfectly absurd, and much that is lamentably true. The view southerly from this point is perhaps better than from most others. We are now looking over a porphyry tract exclusively, where few hard-wooded trees are seen, but pines and cedars, in extensive groves, occupy the whole. From the great ridge, on a southern spur of which we stand, to the level land of the city, a constant diversity appears; bushy and peaty swamps succeed to abrupt elevations, sheltered valleys alternate with projecting rocks and crags. and occasionally we see tokens of a hidden brook that goes down from nook to nook by rapid descent. though there is not, to my knowledge, a waterfall within the territory.

Let me point out to you a few of the more important spots. They will often be referred to in the list that follows. and should your patience survive the ordeal of this introduction, some knowledge of them might not be useless.

Next before us on the southeast, and down. down, many times the "full fathom five" of Ariel, lies Dog Hill Swamp. It is a place where the White Azalea dwells. crowned with especial loveliness: where the Poison Sumach grows to the diameter of a coffee-cup. and woe be to the susceptible woodsman that cuts it; a place where the yellow wasp likes to congregate among the yard-high brakes, and the surveyor works his way in at long intervals, and drives down his stake for a boundary. foot after foot. through the elastic moss. before he strikes hard bottom.

Beyond it is Dog Hill, its namesake and counterpart: where the Bush Clover plumes itself and frolics over the shady sward. On the other side, as in duty bound. another swamp occurs. It is known as Bennet's, and its long trough stretches eastward a full mile. From this and the last. two copious brooks are poured during most of the year, falling into that picturesque sheet of water seen a little to the west of them and called Breed's Pond.

This pond is purely artificial, every foot of it being raised by the dam below. It occupies what was formerly known as Pine Hill Swamp; a labyrinth of bogs and tangled water-bushes, where the otter burrowed in olden time, and round which the wolves went prowling malignantly. On that little rounded island that blends with the southern shore, we could find two deep pits and the remains of another. dug and stoned up by the early farmers as traps for those pests of the flocks. This swamp was stored with the Marsh Marigold or Mayblob before its inundation. and many a basketful have I carried home from thence, crisp and delicate, for a grateful meal in early spring. Pine Hill breaks the prospect beyond the pond, sloping gently down toward the east, as it approaches Bennet's Swamp: and just to the left of those smart-looking buildings that we see yet further off, upon Linwood, we might visit the dilapidated sheep-fold where, long ago, the town shepherd made sure of the nightly safety of his charge. It stands in an angle of the pasture to which Pine Hill has given a name, a crumbling relic of the agricultural age of Lynn, that period now finished, and half forgotten in the age of manufactures.

You see a range of land of somewhat uniform elevation to the west of Breed's Pond, but to the northwest it suddenly sinks to a deep and horse-shoe-formed depression, many acres in extent, and filled by the almost impassable thickets of Tomlin's Swamp. Twenty minutes' walk will bring us to its western border.

Tomlin's Swamp is probably the most extensive morass in this vicinity, with the possible exception of Blood's Swamp. Like the transformed Pine Hill Swamp, it has its tributaries and its outlet, this last being the romantic, though not enduring, Penny-Bridge Brook. The valley by which this stream finds its way to Saugus River is one of the most enchanting seclusions to be found in this district, a place precisely fitted for reverie and contemplation,

> "Where bright green moss heaves in fantastic forms,
> Speckled with sunshine; and but seldom heard,
> The sweet bird's song becomes a hollow sound;
> And the breeze, murmuring indivisibly,
> Preserves its quiet murmur most distinct
> From many a note of many a waterfall,
> And the brook's chatter."

We will turn backwards now, over Penny Bridge again. It is a small, unpretending, rustic affair, rather the worse for neglect and of no manner of importance, save only that one of the boundary stones between Lynn and Saugus stands near it. From here we will stroll away southerly on the western declivity of high hills covered with pines, and presently we find ourselves on the margin of Edwards' Swamp, endowed by Flora more liberally, perhaps, than any of the others. It is the home of the Purple Orchis, and the chosen seat of the Cardinal-flower; the Mayblob is here too, with the Cress, the Golden Senecio and the charming Calla; the Dwarf Cornel and the Pyrola are scattered round the edges, and Fleabanes, Coral-roots, Willow-herbs and Wild Sunflowers are variously distributed on the near hillsides.

From this we will follow the channel of Birch Brook, that flows from its southern extremity. In so doing we pass the remnants of a large cedar-swamp from which many of the trees have been lately cut, and notice the Indian Poke springing greenly by the stream, and the Hempweed climbing in wild freedom over the bushes, both far away from others of their kind. Below this opens the verdant expanse of Pan Swamp Meadow, a rather profitless tract long since reclaimed, that is, tortured out of its natural fitness and character. Birch Brook here unites with that from Breed's Pond, which I have chosen to call Moore's Brook, preferring the name by which it is known in the old deeds of land on its banks, to that of Beaver Brook, sometimes applied to it. As the current passes on through the meadows, it furnishes many a pleasant spot to the botanist, till it falls into Saugus River at the Stone Factory.

I have taken you a long jaunt, kind reader, too bare of interest, perhaps, to be very pleasant. If you desire more minute knowledge of the plants of Lynn, the following list will afford it, I think, very fully. Most of them have passed my own examination. For others, of which I could obtain no specimens, I am glad to inform you of my obligations to Drs. Holder, Clark and Nye, and my friend, Mr. Moulton, of this city; gentlemen keenly alive to the attractions of botany, and possessed of a fund of that pleasant information, scraps of which you will find credited to their several names. For the purpose of connecting our flora with that of the rest of the county, I have interspersed notices of plants unknown here, but detected in other towns. An essential service in this particular has been rendered by the Proceedings of the Essex Institute, issued in 1856, to which society an acknowledgement is also due.

And now I must leave you, reader. There are "tongues in the trees, books in the running brooks, and good in everything." If you love plants, then study them, give yourself to them, and fill your desire with their innocent loveliness. Theirs is display, but not meretricious; beauty, but not heartless and trifling; exciting attractiveness, but no bitter depression follows. To him who reads them aright, they tell a higher tale beside, a tale that none can utter, a tale of life, death, a golden hope and a sanctified immortality. Adieu.

> " The Spring is here ! the delicate-footed May,
> With her slight fingers full of leaves and flowers ;
> And with it comes a thirst to be away,
> Wasting in woodpaths the voluptuous hours.
> A feeling that is like a sense of wings,
> Restless to rise above these perishing things."

LYNN, *May* 1, 1858.

STUDIES OF THE ESSEX FLORA.

POLYPETALOUS EXOGENS.

RANUNCULIDS.

(Crowfoot Family.)

Clematis Virginiana. L.   *Traveller's Joy, Virgin's Bower.*
Common. Low grounds and fence-rows. Cultivated to some extent.

Anemone Virginiana. L.   *Tall Anemone.*
Frequent. Bushy low places. Wood margins.

Anemone nemorosa. L.   *Common Anemone.*
Abundant. Moist, rich ground. Edges of copses.

Hepatica triloba. Chaix.   *Blue Hepatica, Noble Liverwort*
Occasional. Near Pirate's Glen and Howard's Spring, Saugus. Near Dungeon Rock. Woodside, Swampscott.

Thalictrum anemonoides. Michx.   *Rue Anemone.*
Occasional. Fairmount, near Breed's Pond.

T. dioicum. L.   *Early Meadow Rue.*
Occasional. Border of Stacy's Brook, near Essex Street.—*Dr. Holder.*

Large Meadow Rue.	**T. cornuti.** L. Abundant. Meadows and damp places.
Water Crowfoot.	**Ranunculus aquatilis.** L. *Var. divaricatus.* I have found abundant in Saugus River, near Howlet's Mills, Saugus.
Yellow Water Crowfoot.	**R. multifidus.** Ph. I find near Howard Place, Saugus.
Creeping Spearwort.	**R. Flammula.** *Var. reptans.* Grows on the beach at lower end of Middleton Pond.
Seaside Crowfoot.	**R. Cymbalaria.** Pursh. Rare. Nahant.
Small flowered Crowfoot.	**R. abortivus.** L. Occasional. Ledges and rocks in light, rich soil. Dungeon Rock.
Cursed Crowfoot, Celery-leaved Crowfoot.	**R. sceleratus.** L. Rare. Calf Spring, Nahant.
Early Crowfoot.	**R. fascicularis.** Muhl. Front of Common Hill, Swampscott.
Meadow Crowfoot, Creeping Crowfoot.	**R. repens.** L. Abundant. In all low grounds.
Buttercups, Bulbous Crowfoot.	**R. bulbosus.** L. Common. Uplands and fields.
Tall Crowfoot.	**R. acris.** L. Frequent. Moist grass-lands.
Cowslip, Mayblob.	**Caltha palustris.** L. Occasional. Formerly found largely in Edwards' Swamp, now Birch Pond. About Oaklandvale, Saugus.

Coptis trifolia. Salisb.   Goldthread.
Frequent. On the north declivity of Blood's Swamp Hills, it often covers the earth for large spaces.

Aquilegia Canadensis. L.*   Columbine.
Frequent. Rocks and ledges.

Actæa spicata. L. *Var. alba.*   White Cohosh.
Occasional. Pirate's Glen and sparingly in Dungeon Pasture; more frequently in woods, near Upper Swampscott.

MAGNOLIDS.
(MAGNOLIA FAMILY.)

Magnolia glauca. L.   Sweet Bay.
To be found at Gloucester at the well-known locality.

BERBERIDS.
(BARBERRY FAMILY.)

Berberis vulgaris. L.   Barberry.
Abundant. Woods, fields, and almost everywhere.

NYMPHIDS.
(WATER-LILY FAMILY.)

Brasenia peltata, Pursh.   Water-shield.
Frequent. Flax Pond and other waters in the eastern part of the city.

Nymphæa odorata. Ait. †   Pond Lily.
Frequent. Lily Pond and others of its kind.

---

* Easy to cultivate, if good roots be obtained. A fixed variety of a pale, yellowish red or salmon color has been found near Salem. It propagates itself without change.—*Phippen.*

† A curious observation on this plant was made by the late Asa T. Newhall, Esq., of Lynnfield, some years since. Having planted some roots in a small pond fed by a cold spring, the flowers, though luxuriant, were perfectly devoid of odor; a fact due, as he judged, to the coldness of the water where they grew.

Yellow Water-Lily, Cow Lily. **Nuphar advena.** Ait.
Frequent. Ponds and sluggish waters.

SARRACENIDS.

(PITCHER-PLANT FAMILY.)

Side-Saddle Flower, Forefather's Cup, Huntsman's Cup. **Sarracenia purpurea.** L.*
Frequent. Bogs and open swamps.

PAPAVERIDS.

(POPPY FAMILY.)

Celandine. **Chelidonium majus.** L.
Common. Old gardens and cultivated grounds.

Bloodroot. **Sanguinaria Canadensis.** L.
Rare. Moist woods S.E. of Swampscott Cemetery, and also in ravines N.W. of the same point.

FUMARIDS.

(FUMATORY FAMILY.)

Pale Corydalis **Corydalis glauca.** Pursh. †
Frequent. Retired places on ledges, in thin soil.

Common Fumitory. **Fumaria officinalis.** L.
Is sparingly naturalized in N. Andover.

SINAPIDS.

(MUSTARD FAMILY.)

True Water-Cress. **Nasturtium officinale.** R. Br.
Occasional. Brooks and small streams; as for instance, in Neptune Street, near Elm, and near head of Birch Pond.

---

* This fine plant is becoming somewhat classical in American design; and enters into the composition of the fountains in front of the State House, Boston.

† A lovely biennial. By collecting the seeds in mid-summer or the young plants in late autumn, it may be cultivated with ease.

N. palustre. D.C. — Marsh Cress.
Is in the meadows by Shute's Brook, in front of R.R. Station, Saugus Centre.

N. Armoracia. Fries. — Horse-radish.
Often escaped from cultivation.

Cardimine hirsuta. L. — Bitter Cress.
Occasional. Ditches and small streams. Formerly very plentiful in Edwards' Swamp.

Var. VIRGINICA. Michx.
Rare. Retired ledges about Tomlin's Swamp.

Barbarea vulgaris. R. Br. — Winter Cress.
Frequent. Meadows and rich lands.

Sisymbrium officinale. Scop. — Hedge Mustard. Wild Turnip.
Abundant. Cultivated grounds and yards.

Sinapis arvensis. L. — Charlock.
Detected near Salem.

S. nigra. L. — Black Mustard.
Common. Cultivated grounds and waysides.

DRABA VARNA, L., the Common Whitlow Grass, has a single locality in Danvers, first noticed, it is said, by Dr. A. NICHOLS. — *Proc. Ess. Inst.*, 1856. Said to be now extinct.

Lepideum Virginicum. L. — Wild Pepper-Grass.
Common. Roadsides and waste lands.

L. arvense. — Lepidium.
Has been found at Peters' Point, Salem.

Capsella Bursa-pastoris. Mœnch. — Shepherd's Purse.
Abundant. Neglected gardens and compost heaps.

Cakile Americana. Nutt. — Sea Rocket.
Common. Beaches at Nahant, Swampscott, etc.

Raphanus Raphanistrum. L.* — Wild Radish.
Abundant. A garden weed.

---

* A crucifer of unknown species, or at least not included in Gray, has appeared in a single instance at Burrill's Hill.

VIOLIDS.
(Violet Family.)

Lance-Leaved Violet.
Viola lanceolata. L.
Abundant. Meadows and damp fields. Probably our most fragrant species.

Sweet White Violet.
V. blanda. Willd.
Abundant. Bogs and water-courses.

Arrow-Leaved Violet.
V. sagittata. Ait.
Abundant. In every soil and situation. flourishing freely with little regard to circumstances.

Hooded Violet.
V. cucullata. Ait.
Abundant. Meadows and brooksides. A few peculiar specimens I formerly took for *V. palustris*, L.
A cultivated form of this took on, in 1863, the form of *V. palmata*. and the next year reversed to the normal type.—*Phippen.*

Horse Violet. Bird-foot Violet.
V. pedata. L.
Abundant. Dry hills and uplands. The finest species in the region.

Spreading Violet.
V. Muhlenbergii. Torr.
Occasional. In Swamps near Pirate's Glen, Saugus; about Breed's Pond, and so eastward.

Downy Yellow Violet.
V. pubescens. Ait.
Very rare. A rich, shady spot on the back of Bennet's Swamp alone affords it, I believe.

CISTIDS.
(Rock-Rose Family.)

Frostweed, Rock-rose.
Helianthemum Canadense. Michx.
Abundant. Pastures and dry hills and woods.

Hudsonia.
Hudsonia tomentosa. Nutt.
Rare. Short Beach. Nahant.

Large Pinweed
Lechea major. Michx.
Common. Grassy woodlands.

Small Pinweed
L. minor. Lam.
Common. Pastures and clearings.

DROSERIDS.

(SUNDEW FAMILY.)

Drosera rotundifolia. L. — Round-leaved Sundew.
Occasional. Fresh marshes of Chestnut Street, and other such places.

D. longifolia. L. — Long-leaved Sundew.
Occasional. Along the course of Stony Brook, and many similar places.

Parnassia Caroliniana. Michx. — Grass of Parnassus, Whiteblow.
Very rare. I have found it in Lynnfield, and along the banks of Stacy's Brook.

HYPERIDS.

(ST. JOHN'S-WORT FAMILY.)

Hypericum perforatum. L. — St. John's-wort.
Abundant. Roadsides and neglected soils.

H. mutilum. L. — Small St. John's-wort.
Common. A weed in fields and gardens.

H. Canadense. L. — Canadian St. John's-wort.
Common. Gravelly edges of ponds and watercourses.

H. Sarothra. Michx. — Pineweed.
Abundant. Dry uplands, in cartways and excavations.

Elodea Virginica. Nutt. — Marsh St. John's-wort.
Abundant. Cool and shady bogs and along sluggish waters.

ELATINIDS.

(WATER-WORT FAMILY.)

Elatine Americana. Arnott. — Water-wort.
Rare. Flax Pond.—*Robinson.*

DIANTHIDS.
(Pink Family.)

Wild Pink, Deptford Pink.
Dianthus Armeria. L.
Frequent. Turfy banks, mostly near roads.

Bouncing Bet, Soapwort.
Saponaria Officinalis. L.
Common. Old grounds about houses.

Bladder Campion, Crackers.
Silene inflata. Smith.
Abundant. Gravelly places and roadsides. Remarkable for its main root, which is sometimes enormously long.*

Sweet William, Catchfly.
S. Armeria. L.
Frequent. A garden weed often migrating to the street.

Wooly Pink, Corn Cockle.
Lychnis Githago. Lam.
Occasional. Gardens and fields.

Thyme-leaved Sandwort.
Arenaria serpyllifolia. L.
Occasional. Damp rocks at Pine Hill and old cultivated spots about Atlantic St. A little in doubt.

Side-Flowering Sandwort.
A. lateriflora. L.
Abundant on bushy hillocks between the dyke and the railroad, and in other like places. Mudge's Woods. Swampscott.
Borders of ponds about Legg's Hill. Salem.— *Phippen.*

Sea Sandwort.
Honkenya peploides. Ehr.
Frequent. King's Beach, and others in that region.—*Dr. Holder.*

Chickweed.
Stellaria media. Smith.
Abundant. Gardens and fields.

Northern Stitchwort.
S. borealis. Big.
Has been detected in Broad Meadow. Lynnfield. —*A. P. Chute.*

Mouse-ear Chickweed.
Cerastium Viscosum. L.
Abundant. Cultivated grounds and waysides.

---

* I traced one five feet into the earth perpendicularly, which retained a diameter of nearly an inch. This plant is almost unknown in Salem, and the north part of the county.—*Phippen.*

C. arvense. L. — Field Chickweed.
Very rare in Lynn, if here at all. Abundant at Nahant, whitening the pastures when in flower, and sparingly at Swampscott on the shore ledges.

Sagina procumbens. L. — Pearlwort.
Rare. Spring ledges near Pirate's Glen, Saugus. Occasional on gravelly bluffs on the east side of Nahant. Also at Rockport, 1862.

Spergula arvensis. L. — Corn Spurry.
Common. Waste places in cultivated grounds, stubble fields.

Spergularia rubra. Pers. — Sandwort.
Common. Waysides and yards.

Var. MARINA.
Frequent on Commercial wharf and elsewhere.

Scleranthus annuus. L. — Knawel.
Common. Dry gravelly fields and roadsides.

Mollugo verticillata. L. — Carpet weed.
Common. Gardens and fields in rich, cultivated soil.

PORTULIDS.
(PURSLANE FAMILY.)

Portulaca oleracea. L. — Purslane.
Abundant. No garden can be long tilled without producing it. Thought to be naturalized from Europe, but a French writer in 1636 says: "Purslane naturally comes to the Indians in their cultivated fields, among their corn and pumpkins and is common with them."—*Phippen*.

MALVIDS.
(MALLOW FAMILY.)

Abutilon avicennæ. Gært. — Velvet-leaf. Wild Cotton.
Frequent. It occurs spontaneously in gardens and manured lands.

Low Mallows, Cheesevine. **Malva rotundifolia.** L.
Abundant. Everywhere near houses and buildings; well known.

TILIDS.

(LINDEN FAMILY.)

American Linden, Basswood. **Tilia Americana.** L.
Occasional. I have found it in Dungeon Pasture: also on Second Pine Hill. It is more plentiful at Oak Island, Chelsea.

LINIDS.

(FLAX FAMILY.)

Wild Flax. **Linum Virginianum.** L.
Very rare. I have only once met with it, near the top of Second Pine Hill where it still flourishes in small quantity.

GERANIDS.

(GERANIUM FAMILY.)

Cranesbill, Wild Geranium. **Geranium maculatum.** L.
Frequent. Thickets, borders of swamps and retired fence rows. Very readily cultivated, yielding a profusion of beautiful flowers. The curious projection of its ripe seeds is worthy of study.

Carolina Cranesbill. **G. Carolinianum.** L.
Occasional. Cultivated lands, not elegant.

Herb Robert. **G. Robertianum.** L.
Abundant. Rocky places, at the base of ledges, flourishing among the debris. Miserably fetid.

Touch-me-not. **Impatiens pallida.** Nutt.
At Backside, Hamilton, side by side with *I. fulva.—Phippen.*

Jewel-weed. **I. fulva.** Nutt.
Abundant. Brooksides and wet, rich quagmires.

Oxalis stricta. L.   Yellow Wood Sorrel, Ladies' Sorrel.
Common. In all soils not too dry, particularly manure-beds and other rich places.

ANACARIDS.

(CASHEW FAMILY.)

Rhus typhina. L.*   Staghorn Sumach.
Abundant. Most common on hills in light soil, but thrives almost anywhere. Difficult to eradicate.

R. glabra. L.   Smooth Sumach.
Abundant. Intermixed with the former.

R. copallina. L.   Dwarf Sumach.
Frequent. Hillsides in good soil, often forming crowded patches.

R. venenata. DC.   Poison Elder, Dogwood, Poison Sumach.
Frequent. Swamps and wet thickets. Its poisonous property affects different persons in various degrees, many not being susceptible at all.

R. Toxicodendron. L.†   Poison Oak, Mercury, Poison Ivy.
Common. Not a respecter of soils by any means.

VITIDS.

(VINE FAMILY.)

Vitis Labrusca. L.   Common Wild Grape.
Frequent. Damp thickets. The fertile plants comparatively few.

---

* In this vicinity, this and the two following show to rather poor advantage in autumn, being divested of all the young shoots and leaves by the morocco tanners. *R. copallina* is said to be better for this than the others, though less abundant.

† It is a terror to many who are not poisoned by *R. venenata*, of which class I happen to be one.

A marked specimen of the shrub-form is at Juniper Point, Salem.—*Phippen*.

Summer Grape.
V. æstivalis. Michx.
 Occasional. It seems to occupy higher situations than *V. Labrusca.* and to be more generally fruitful. *V. Cordifolia.* Michx.. the Frost Grape. I think is with us but cannot be at all positive.

Creeper, Woodbine.
Ampelopsis quinquefolia. Michx.
 Abundant. Thrives in all kinds of soil where there is not too much shade.

RHAMNIDS.
(BUCKTHORN FAMILY.)

Jersey Tea.
Ceanothus Americanus. L.
 Occasional. The best localities are on the southern slope of Linwood. and a hillside west of Sadler's Rock.

CELASTRIDS.
(SPINDLE-TREE FAMILY.)

Staff-tree. Waxwork. Bitter Sweet.
Celastrus scandeus. L.*
 Abundant. Borders of fields. and fencerows. A luxuriant and elegant vine. and easy of cultivation.

SAPINIDS.
(SOAPBERRY TREE.)

Striped Maple.
Acer Pennsylvanicum. L.
 Rare. Said to grow in Dungeon Pasture.

Sugar Maple.
Acer saccharinum. L.
 Occasional. I have not found it in Lynn. although it is probably here. It grows at Woodside. Swampscott.

Swamp Maple.
A. rubrum. L.
 Abundant. The most common tree in swamps and wet localities, next the Alder.

---

* Not only graceful. but curiously illustrative of the tendency of leaves to turn the upper surface to the light: the petioles being all twisted when the twig is inverted.

POLYGALIDS.
(MILKWORT FAMILY.)

Polygala sanguinea. L.* *Red Milkwort.*
Common. A showy little plant in meadows and damp lands, flowering after the grass is mown. The root has a fine odor like that of Checkerberry.

P. cruciata. L. *Cross-leaved Milkwort.*
Very rare. I have met with it in wet, oozy lands below the mill-dam of E. Holmes, Lynn. The best locality I know of is Marblehead Great Neck, where it is abundant in the damp lands.

P. verticillata. L. *Whorled Milkwort.*
Occasional. I know no localities except a meadow near the head of Raddin's Ct., West Lynn, and a spot on the highlands of Rocks Pasture.

P. polygama. Walt. *Double-fruited Milkwort.*
Frequent. Most so upon and near Second Pine Hill. A fine plant in flower.

PISIDS.
(PULSE FAMILY.)

Lupinus perennis. L. *Lupine.*
Said to grow in Lynnfield.

Lathyrus maritimus. Bigel. *Beach Pea.*
Abundant. Stony beaches of Nahant and elsewhere.

L. palustris. L. *Marsh Vetchling. Meadow Pea.*
Rare. Confined entirely to the Eastern section. Borders of Stacy's Brook, near Humphrey Street.

Apios tuberosa. Mœnch. *Ground-nut.*
Abundant. Overrunning bushes in damp thickets, and spreading extensively.

Amphicarpæa monoica. Nutt. *Hog Pea-nut.*
Frequent. Shady places in light, moist soil.

---

* This genus is a difficult study for beginners, but is perfectly easy to recognize after a little acquaintance.

Naked flowered Tick-Trefoil.	**Desmodium nudiflorum.** DC.	

Rare. It grows scantily on the southern side of Blood's Swamp Hills; elsewhere I have not seen it.

Desmodium. **D. acuminatum.** DC.

A ledge on Blood's Swamp Hills, nearly north of the head of Dog Hill Swamp furnishes this plant. No other locality is known to me. (July, 1862.)

Tick-Trefoil. **D. Canadense.** DC.

Is occasional on the railroad at that part of the city formerly known as Woodend, and along the hillsides west of Sadler's Rock.

Bush Clover. **Lespedeza violacea.** Pers.

Frequent. Dry oak and hickory woods. A handsome plant, and well worth cultivating.

Hairy Bush Clover. **L. hirta.** Ell.

Common. Open, gravelly hillsides and fields.

Wooly-stemmed Clover. **L. capitata.** Michx.

May be determined among the numerous forms found on Linwood, and other like places.

Common Locust. **Robinia pseudacacia.** L.

Occasional. Sparingly naturalized. Rather freely established at Cider Mill Pasture, west of Pan Swamp Meadow. Abundant in Peabody, near Lynn line.

Rabbit-foot Clover, Pussy Clover. **Trifolium arvense.** L.

Abundant. Dry, sterile fields and roadsides.

Red Clover. **T. pratense.** L.

Abundant. Cultivated largely and thoroughly naturalized here. A variety with pure white flowers is occasional in Saugus, Lynn, Salem, etc.

Zigzag Clover. **T. medium.** L.

Found in Danvers, near Topsfield line.

White Clover, Honeysuckle. **T. repens.** L.

Abundant. Establishing itself almost everywhere, resisting drought and flourishing in the most unpropitious seasons.

T. agrarium. L.   Hop Clover.
Common in poor grass land about Salem.—*Phippen*.

T. procumbens. L.   Low Hop Clover.
Rare. I have only found it in Linnean Grove. A curious little species.

Melilotus officinalis. Willd.   Sweet Clover.
Is on the beaches at Cohasset Rocks. In cultivated clover, Salem Neck.—*Phippen*.

M. leucantha. Koch.   White Melilot.
Is naturalized in Rowley.—*Proc. Ess. Inst.*, 1856.

Medicago lupulina. L.   Nonesuch.
Common. Gardens and fields. Also frequently found by waysides and about houses.

Genista tinctoria. L.   Wood Waxen, Dyer's Weed.
Abundant. Between Lynn and Salem is no doubt the chief seat of this pernicious intruder. It completely covers the elevated pasture soil, exterminating the grass and almost every plant beside.

Baptisia tinctoria. R. Br.   Wild Indigo.
Abundant. Pastures and hills; troublesome in tilling new lands.

ROSIDS.

(ROSE FAMILY.)

Prunus maritima. Wang.   Beach Plum.
Occasional. Singularly enough, not on the shore at all, but along Boston Street, near Wyoma, and like places. Frequent about Nahant, on the bluffs.

P. Pennsylvanica. Loisel.   Wild Red Cherry.
Frequent. Hillsides and pastures. An elegant species.

P. Virginiana. DC.   Choke Cherry.
Common. Upland thickets and old fence-rows.

Black Cherry.	**P. seratina.** DC.	

Common. Woods. The best specimens in Lynn are probably in the old Western Burial Ground, Market Square.

Meadow-sweet. **Spiræa salicifolia.** L.
Abundant. Swamps and boggy water-courses.

Hardhack. **S. tomentosa.** L.
Abundant. Rich pastures and uplands. Too common to be duly appreciated.

Agrimony. **Agrimonia Eupatoria.** L.
Common. Thickets and borders of swamps.

White Avens. **Geum Virginianum.** L.
Occasional. Shady, rich soil among rocks in elevated spots.

Tall Yellow Avens. **G. strictum.** Ait.
Rare. Formerly at roadside at North Bend.

Purple Avens. **G. rivale.** L.
Rare. Hardly to be found except in Linnean Grove.—*Dr. Holder.*
Topsfield. Formerly at N. Salem.—*Phippen.*

Norway Cinque-foil. **Potentilla Norvegica.** L.
Frequent. Cultivated grounds and around dwellings.

Five-finger. **P. Canadensis.** L.
Abundant. *Var. pumilla* is a pioneer plant in almost every soil; *Var. simplex* occurs frequently along walks and in dry thickets.

Silvery Cinque-foil. **P. argentea.** L.
Common. Borders of streets and travelled ways. Remarkable for its downy whiteness.

Silver-weed. **P. anserina.** L.
Abundant. Dykes and banks about the salt marshes. Resembles tansy.

Crowded Cinque-foil. **P. arguta.** Pursh.
Rare. Formerly at Burrill's Hill.
Sunny cliffs about Castle Hill, S. Salem. *Buttrick.*

P. fruticosa. L. — Shrubby Cinque-foil.
Very rare in Lynn, if here at all. I have it from peat bogs in Lynnfield, near Serpentine Quarry, where it abounds. " Backside." Hamilton.—*Phippen.*

P. tridentata. Ait. — Mountain Cinque-foil.
Near Bass Rock, Gloucester.—*J. L. Russell,* 1861.

Fragaria Virginiana. Ehrh. — Strawberry.
Common. Meadows and fields. Plentiful in recent clearings.

F. vesca. L. — Long-fruited Strawberry.
Frequent. Low grounds. Both species bear very scantily in this section.

Rubus odoratus. L. — Flowering Raspberry.
Is said to be naturalized in Salem Pastures.

R. strigosus. Michx. — Red Raspberry.
Abundant. Rocky places in rich woodlands; the fruit generally small in quantity.

R. occidentalis. L. — Thimbleberry.
Common. Along pasture walls and in the edges of thickets. Generally very fruitful.

R. villosus. Ait. — High Blackberry.
Abundant. Damp soils generally; the varieties appear very much intermixed. The fruit for the most part is poor and valueless.

R. Canadensis. L. — Low Blackberry.
Abundant. Open pastures and hillsides, trailing extensively and bearing great quantities of fruit.

R. hispidus. L. — Swamp Blackberry.
Abundant. Filling every swamp, and thriving almost as well on shady uplands. Fruit of little consequence.

Rosa lucida. Ehrh. * — Low Wild Rose.
Common. Mostly on the edges of swamps and in damp thickets.

---

* Either this and R. *Carolina,* L., the Swamp Rose, are not well distinguished, or else the latter is not in this region.

Sweetbrier.	**R. rubiginosa.** L.	

    Frequent. Chiefly in open hillsides and uplands.

Red Thorn.   **Cratægus coccinea.** L.

    Rare. I found it formerly at Burrill's Hill and in one or two other spots. Fine specimens may be seen in Salem, near the crossing of the Marblehead Railroad and the Forest River road.

Chokeberry, Dogberry.   **Pyrus arbutifolia.** L.

    Common. Among huckleberry bushes. Fruit profuse and attractive to the eye, but not by any means to the taste.

Mountain Ash.   **P. Americana.** DC.

    Occasional. Small specimens are only to be found, as the trees are early seized upon for cultivation.

Shad-bush, June-berry.   **Amelanchiar Canadensis.** Torr. & Gray.

    Common in almost every low ground.

    Var. BOTRYAPIUM occurs at Lantern Hill, and occasionally in other places. A peculiar form occurs along the shore at Norman's Woe, Gloucester, fruiting plentifully when the bushes are not three feet high.

MELASTOMIDS.

(MELASTOMA FAMILY.)

Meadow Beauty.   **Rhexia Virginica.** L.

    Frequent. Brooksides and meadows. Generally plentiful where it appears at all.

LYTHRIDS.

(LOOSE-STRIFE FAMILY.)

    AMMANNIA HUMILIS. Michx., a small weed with no other name, is said to grow in Danvers.—*Proc. Ess. Inst.*, 1856.

Low Loosestrife.   **Lythrum hyssopifolia.** L.

    I found it at Flax Pond, and also in Oak street, in 1849. I have seen it quite plentifully in a meadow at Nahant, and in lesser quantity in low grounds along Moore's Brook above Boston Street.

Nesæa verticillata. Ell. — Swamp Loosestrife.
Frequent. Ponds and wet swamps. The stems are sometimes curiously thickened under water.

ONAGRIDS.
(FUCHSIA FAMILY.)

Epilobium angustifolium. L. — Great Willow-herb.
Abundant. It seems partial to burnt lands, as the hillside N.E. of Breed's Pond, where in 1856 its flowers made one sheet of purple. Beautiful and easy to cultivate.

E. palustre. L. — Epilobium.
Not rare. Hamilton, Middleton and Danvers.— *Phippen.*

E. coloratum. Muhl. — Purple-veined Willow-herb.
Frequent. Springy spots and borders of ponds.

Œnothera biennis. L. — Evening Primrose, Scabish.
Common. Gardens and fields. I have found it in Rock's Pasture, with flowers more than double the common size and very showy.

Var. CRUCIATA is very abundant on Eastern Railroad between Hamilton and Ipswich.—*Phippen.*

Œ. pumila. L. — Dwarf Evening Primrose.
Frequent in dry gravelly spots, and sometimes in moist grounds.

Ludwigea alternifolia. L. — Seed-box.
Rare. Formerly on Washington Street, between Essex and Laighton. I found it nowhere else and then it was not plentiful.

L. palustris. Ell. — Water Purslane.
Abundant. Creeping in the mud on the banks of ponds and stagnant waters.

Circæa Lutetiana. L. — Enchanter's Nightshade.
Occasional. Dungeon Pasture and elsewhere. Plants seldom solitary.

C. alpina. L. — Small Enchanter's Nightshade.
Occasional. Pirate's Glen. Partial to shady spots and grows in patches.

Mermaid Weed. **Proserpinaca palustris.** L.

In most ponds and permanent ditches; generally in shallows that dry in summer.

Variable Water-Milfoil. **Myriophyllum ambiguum.** Nutt.

Var. MATANS. Occasional. Breed's Pond. Probably in many still waters.

Var. CAPILLACEUM. Occasional. In a rocky pond-hole in Marshall's Pasture.

CACTIDS.

(CACTUS FAMILY.)

Prickly Pear. **Opuntia vulgaris.** Mill.

On the Ipswich River bank, at North Reading, where a few plants were placed many years ago there is now a very flourishing locality.—*Robinson.*

CRASSULIDS.

(HOUSELEEK FAMILY.)

Mossy Stone-crop, Golden Moss. **Sedum acre.** L.

Occasional. Naturalized between the old hotel and Spouting Horn, Nahant, on the gravel banks.

Aaron's Rod, Live-forever. **S. Telephium.** L.

Occasional. Near old houses and about ledges.

Ditch Stone-crop. **Penthorum sedoides.** L.

Common. Muddy ditches and edges of stagnant pools. I have never seen the petaloid state.

Houseleek. **Sempervivum tectorum.** L. (?)

Rare. Naturalized on a ledge at E. Stone's house on Boston Street. Also on a rock at Oaklandvale, Saugus, and on rocks at south end of Floating Bridge Pond.

On rocks, near Witch Hill, Salem —*Phippen.*

SAXAFRIDS.

(SAXAFRAGE FAMILY.)

Saxifraga Virginiensis. Michx.   Mousemead, Early Saxifrage

Common. Wherever there is a damp rock to sustain it, it may be found full of buds as soon as the snow is gone.
Var. CHLORANTHA, with green flowers, occurs in Topsfield.—*Proc. Ess. Inst.*, 1856.

S. Pennsylvanica. L.   Swamp Saxifrage.

Rare. Perhaps not in Lynn at all, but plentiful in the wet edges of Shute's Brook, near railroad station, Saugus Centre.
In a swamp in Great Pasture, Salem, and along Ipswich River, Hamilton. Roadsides in Wenham, and Rocks near Beverly Bridge, Salem. Abundant and very large along roads in Topsfield, near Ipswich River.—*Phippen.*

Tiarella cordifolia. L.   False Mitrewort.

Rare. It has been found within the township.—*Dr. Holder.*

Chrysosplenium Americanum. Schw.   Golden Saxifrage.

Occasional. Saugus and Swampscott.

Ribes hirtellum. Michx.   Short-stalked Gooseberry.

Abundant. On almost every rocky hill, and by no means rare in low grounds. Generally quite fruitful.

HAMAMELIDS.

(WITCH-HAZEL FAMILY.)

Hamamelis Virginica. L.   Witch Hazel.

Abundant. Damp hillsides, near swamps; easily known by its being in full flower when there are no leaves to conceal it from sight.

APIDS.

(PARSLEY FAMILY.)

Hydrocotyle Americana. L.   Pennywort.

Abundant. Every swamp is filled with it.

H. UMBELLATA, L., (?) the Round-leaved Pennywort, I have found at Essex Pond, Hamilton, but not in flower, so that I rather doubt the species.

Crantzia. **Crantzia lineata.** Nutt.
A little plant not otherwise named grows on the brackish marshes at Salisbury.—*A. P. Chute.*

Sanicle. **Sanicula Marilandica.** L.
Frequent. Edges of thickets and among low bushes.

Carrot. **Daucus Carota.** L.
Occasional. Naturalized in old fields.
HERACLEUM LANATUM, Michx.
The Cow Parsnip is probably not in Lynn, but grows on the Forest River Road.

Parsnip. **Pastinaca sativa.** L.
Occasional. Plentifully established in some older parts of the city.
ARCHANGELICA PERIGRINA. Nutt.
The lesser Angelica is to be found in Salem, Danvers and Beverly.—*Proc. Ess. Inst.*

Fool's Parsley. **Æthusa Cynapium.** L.
Occasional. In old gardens and cultivated fields.

Scotch Lovage. **Ligusticum Scoticum.** L.
Frequent. Perhaps not in the proper territory of Lynn, but scattered liberally along the shores about Nahant.

Meadow Parsnip. **Thaspium aureum.** Nutt.
I have found in Topsfield, near Boxford.

Musquash-Root **Cicuta maculata.** L.
Frequent. In swamps and like places.

Bulb-bearing Water Hemlock **C. bulbifera.** L.
Occasional. Edges of Strawberry Brook, in the Bowler Swamp, and elsewhere.

Water Parsnip. **Sium latifolium.** L. (?)
Frequent. Brooks and swamps. The distinction between this and *S. lineare*, Michx., is too obscure.

OSMORRHIZA BREVISTYLIS, DC., the Hairy Sweet Cicely, grows, it is said, at Oak Island, Chelsea, but I doubt its being an inhabitant of Lynn.

CANIUM MACULATUM, L., was formerly in the streets of Salem, but it has disappeared.—*Phippen.*

CARUM CARUI, L., the common Caraway, is naturalized in Rowley and Ipswich.

ARALIDS.
(SPIKENARD FAMILY.)

Aralia racemosa. L.   Spikenard, Pettimorril.
Rare. I have found but one specimen in the place, which was in the valley S.E. from Dungeon Rock. It has, however, been brought from the woods for cultivation by others.

A. nudicaulis. L.   Wild Sarsaparilla.
Abundant. Rocky hills and woods.

A. hispida. Michx.   Bristly Sarsaparilla.
Frequent. Hilly pastures. Particularly abundant near Gravel and Round Ponds, Hamilton.

Panax trifolium. L.   Dwarf Ginsing.
Is to be found at the Aqueduct Fountains, Danvers.

CORNIDS.
(CORNEL FAMILY.)

Cornus circinata. L'Her.   Round-leaved Cornel.
Frequent. Grows largely among the rocks on the N.E. slope of Second Pine Hill, near the road.

C. stolenifera. Michx. (?)   Red-osier Cornel.
Frequent. Upper Swampscott.

C. paniculata. L'Her.   Panicled Cornel.
Frequent. Uplands and hillsides in cool soil. I have some doubt of this species being correctly determined.

C. florida. L.   Flowering Dogwood.
Rare. Three or four localities are known.

Dwarf Cornel. **C. Canadensis.** L.

Frequent. Around the margin of Breed's Pond, and elsewhere in Dungeon Pasture.

Alternate-leaved Cornel. **C. alternifolia.** L.

Is frequent on rocky hills and occasional in cultivation.

Tupelo, Pepperidge. **Nyssa multiflora.** Wang.

Abundant. Rich woods and thickets. Very fine in and about Pine Grove Cemetery.

ADDENDA.

TURRITUS GLABRA, L., the Smooth Tower-Mustard, is at Paradise and Orne's Point, Salem.—*Proc. Ess. Inst.*, 1856.

DRABA CAROLINIANA, Walt., the Whitlow Grass, has been found in Salem.—*Ibid.*

SANGUISORBA CANADENSIS, L., the Canada Burnet, may be often found in Hamilton and vicinity, though never, I think, west of Salem.

MONOPETALOUS EXOGENS.

CAPRIFOLIDS.

(HONEYSUCKLE FAMILY)

Linnea, Twin-flower. **Linnea borealis.** Gronov.

Occasional. Very luxuriant on a hill near the Lynnfield road, Wyoma.

LONICERA SEMPERVIRENS, Ait., the Trumpet Honeysuckle has been detected near Marblehead. —*J. L. Russell.*—*Proc. Ess. Inst.*, 1856.

Diervilla trifida. Mœnch.   Bush Honey-suckle.
Frequent. Along walls near the Saugus line, north of Boston Street.
Abundant in Beverly, Manchester and Essex.—*Phippen.*

Triosteum perfoliatum. L.   Feverwort.
Frequent. Cool and moist lands.

Sambucus Canadensis. L.   Elder.
Common. Damp thickets and swamps.

S. pubens. Mx.   Red-berried Elder.
Abundant at Hamilton Ponds.
Has been found at Danvers.—*Russell.*

Viburnum nudum. L.   Withe-rod.
Grows at the head of Middleton Pond, and at Penny Bridge, Saugus.

V. Lentago. L.   Sweet Viburnum.
Occasional. In a pasture west of Pan Swamp Meadow.

V. dentatum. L.   Arrow-wood.
Common. Thickets and woods.

V. acerifolium. L.*   Maple-leaved Arrow-wood.
Occasional. Generally on elevated and fertile soils. Seldom abundant in any place.

RUBIDS.

(MADDER FAMILY.)

Galium asprellum. Michx.   Rough Cleavers.
Abundant. Swamp thickets covering the bushes.

G. trifidum. L.   Small Bedstraw
Abundant. Moist grounds and borders of streams.

G. triflorum. Michx.   Sweet Bedstraw
Frequent. Boggy open places, among ferns and the like. A rather showy species.

---

* An elegant undershrub, well suited for ornamental uses. When not in flower, it is not readily distinguished from a young maple.

Wild Licorice.	**G. circæzans.** Michx. Common. May be found on the south side of almost every precipice where there is sufficient soil.
Button-bush.	**Cephalanthus occidentalis.** L. Abundant. Water-courses and ponds, forming dense jungle-like masses. Its habit reminds one of the tropical Mangrove.
Partridge-berry, Box-berry.	**Mitchella repens.** L. Common. Certain to be in all woods, and often where there are none, yet better worth cultivating than many exotics.
Bluets, Innocence.	**Houstonia cœrulea.** L. Common. In all kinds of grass-lands, except the dryest. Find both living and dead stems on same root, indicating a perennial habit. Doubt its being a biennial.—*Phippen.*

ASTERIDS.

(ASTER FAMILY.)

Blazing Star.	**Liatris scariosa.** Willd. Very rare. Said to grow on the northeast side of Humphrey's Pond, Lynnfield. Reading and Hamilton.—*Phippen.*
Trumpet Weed, Queen o' the Meadow, Indian Hemp.	**Eupatorium purpureum.** L. Abundant. In every swamp and low meadow. Rarely solitary.
Verbena-leaved Boneset.	**E. teucrifolium.** Willd. Occasional. In Bowler Swamp, and also in the Ox Pasture, near Lynnfield line. Backside, Hamilton.—*Phippen.*
Broad-leaved Boneset.	**E. pubescens.** Muhl. Rare. Somewhat plentiful in Swampscott.
Smooth Boneset.	**E. sessilifolium.** L. Rare. Formerly at Burrill's Hill.
Thoroughwort.	**E. perfoliatum.** L. Frequent. Damp situations; not apt to be plentiful, but sometimes abundant.

Mikania scandens. L. — Climbing Hemp-weed.

Very rare. Formerly on the shore of Breed's Pond, at the outlet of Bennet's Brook.
Plenty in swamp at W. Danvers. Very luxuriant along Ipswich River, Hamilton.—*Phippen.*

Sericocarpus conyzoides. Nees. — White-topped Aster.

Common. Pastures and hills.

Aster corymbosus. Ait.* — Corymbed Aster.

Frequent. Shady, moist places. Easy to recognize. A form is now and then found which may be *A. macrophyllus*, L.

A. patens. Ait. — Spreading Aster.

Frequent. Scattered among the berry bushes on warm upland slopes. One of our finest species.

A. lævis. L. — Smooth Blue Aster.

Occasional. Eastern part of the town; also in Swampscott.

A. undulatus. L. — Variable Aster.

Frequent. The most common species in rocky spots, as the east side of Hathorne's Hill. Very conspicuous.

A. cordifolius. L. — Heart-leaved Aster.

Frequent. Borders of brooks and like moist localities.

A. dumosus. L. — Bushy Aster.

Common. More generally distributed than other species, but rather less frequent near dwellings.

A. Tradescanti. L. — Narrow-leaved Aster.

Abundant. Principally by roadsides and in gravelly soils. Noticeable for its close, cylindrical-looking racemes of white flowers.

---

* Not a few of this genus might be made to adorn the garden as much as the Cinerarias. *A. lævis, patens* and *Nova Angliæ*, are all charming flowers, with scarce a fault; but, alas! they are "so common."

Mean Aster.

A. miser. L.

Common. Most readily distinguished by the leaves. The last three species, though wholly distinct, are a difficult study for beginners.

Willow-leaved Aster.

A. longifolius. Lam.

Abundant. The prevailing species in swampy places and along water-courses. Variable. particularly in tint. A well-marked variety of this, as I believe, occurs along the beaches in Swampscott. It has a stiff, bushy aspect and quite obtuse leaves.

Rough stemmed Aster.

A. puniceus. L.

Frequent. The roughest species I have found. Resembles *A. longifolius* in the flowers.

New England Aster.

A. Novæ-Angliæ. L.

Frequent Along the railroad in Ward Seven, and on the West Lynn marshes.

A fine variety with rose-purple flowers. A. ROSENS, Desf., has been shown as gathered in the township.

Pointed-leaved Aster.

A. acuminatus. Michx.

Occasional. Borders of swamps and like situations.

Annual Salt Marsh Aster.

A. linifolius. L.

Frequent. Marshes at West Lynn Station and elsewhere.

Rare or unknown about Salem.—*Phippen*.

Horseweed, Fleabane.

Erigeron Canadense. L.

Common. The inseparable companion of agriculture, showing itself in every garden and field.

Robin's Plantain.

E. bellidifolium. Muhl.

Occasional. Roadsides.

Purple Fleabane.

E. Philadelphicum. L.

Occasional. Near Orne's Point, Salem. Also in Danvers.—*Phippen*.

Daisy Fleabane.

E. annuum. Pers

Frequent. Along Strawberry Brook.

E. strigosum. Muhl. — Narrow-leaved Daisy Fleabean.
Common. Scattered over every pasture and hillside during most of the summer.

Diplopappus linariifolius. Hook. (?) — Violet Diplopappus.
Abundant. Pine Hill and other bushy pastures. I cannot suppose the plant to be anything but this, but the short pappus is altogether obscure. Otherwise the description perfectly applies.

D. umbellatus. Torr and Gr. (?) — Large Diplopappus.
Frequent. Fence rows and elsewhere. Very plentiful near Beck's Pond, Hamilton. The same uncertainty attends this as the preceding.

D. cornifolius. Darl. (?) — Cornel-leaved Diplopappus.
Occasional. Margins of Dog Hill Swamp. Also, in profusion, on the hills S. of Peabody Almshouse.

Solidago bicolor. L. — White-rayed Goldenrod.
Frequent. A pasture plant, here and there among the huckleberries.

S. cæsia. L. — Blue-stemmed Goldenrod.
Abundant. Most frequent in the woods or recent clearings, and in rocky locations.

S. puberula. Nutt. — Many-flowered Goldenrod.
Rare. I have found it in a deserted stone-quarry near the N.W. corner of Pine Grove Cemetery. Also in the reclaimed fields of Stocker's Swamp, Saugus.

S. stricta. Ait. — Willow-leaved Goldenrod.
Occasional. Bowler Swamp and similar places.

S. sempervirens. L. — Seaside Goldenrod.
Abundant. On the dykes and other places around the borders of the marshes.
Makes a fine garden plant, much frequented by the butterflies.—*Phippen.*

S. neglecta. T. & G. (?) — Smooth Goldenrod.
Occasional. Swamps in the neighborhood of Stony Brook.

S. linoides. Solander. — Slender Goldenrod.
Occasional. So far only in the swamp east of Horse Pasture Spring.

Rough or Tall Goldenrod.	**S. altissima.** L.
Common. Fence-rows in old grounds where it is almost constant. Our most conspicuous, though not most elegant species.
Sweet Goldenrod.	**S. odora.** Ait.
Is about Cutler's Pond, Hamilton; and said to be in Lynn.
Gray Goldenrod.	**S. nemoralis.** Ait.
Common. Dry fields and hillsides; the most prominent flower in the autumnal scenery.
Common Three-ribbed Goldenrod.	**S. Canadensis.** L.
Frequent. Dry and gravelly soils.
Late Three-ribbed Goldenrod.	**S. serotina.** Ait. (?)
Occasional. Dry and moist lands, and margins of swamps.
Bushy Goldenrod.	**S. lanceolata.** L.
Frequent. Alluvial soils. The only species with us which has an agreeable odor.
Elecampane.	**Inula Helenium.** L.
Rare. Escaped from cultivation. Along Western Ave. In quiet spots in Saugus Center.
High-water Shrub.	**Iva frutescens.** L.
Rare. Only appears on the old dyke at Willis' Neck.
Salt Marsh Fleabane.	**Pluchea camphorata.** DC.
Grows on the marsh above Bear Pond, Nahant.
Roman Wormwood.	**Ambrosia artemisiæfolia.** L.
Common. Too well-known for any further notice.
Sea Burdock, Cockle-burr.	**Xanthium echinatum.** Murray.
Frequent. On shores and beaches, at about high-water mark. Also inland, growing freely on the beds of sea-manure heaps.
Cut-leaved Rudbeckia.	**Rudbeckia laciniata.** L.
Said to be wild in Lynnfield.

R. hirta. L. — Cone-flower.
Occasional. It appears now and then in the meadows and mowings, but seems inconstant as to locality.

Helianthus divaricatus. L. — Cross-leaved Sunflower.
Frequent. Readily found at Stone Factory, Boston Street; also at Linwood.

H. strumosus. L. — Pale-leaved Sunflower.
Frequent. Only in the eastern section, near Swampscott. The prevailing form does not fully agree with this species, and some might incline to call it *H. decapetalus*.

H. tuberosus. L. — Jerusalem Artichoke.
Frequent. Getting established in by-places in good soil.

Coreopsis trichosperma. Michx. — Tickseed Sunflower.
Occasional. Confined entirely to the eastern part of the city, beginning about Cedar Pond. Low, damp, and marshy spots.

Bidens frondosa. L. — Cuckoldweed, Beggar-ticks.
Common. Cultivated grounds; sure to invite attention at seeding-time.

B. connata. Muhl.* — Swamp Beggar-ticks.
Frequent. Borders of ponds and streams.

B. cernua. L. — Nodding Burr-Marigold.
Occasional. Border of Moore's Brook, above Boston Street, afforded good specimens.

B. Beckii. Torr. — Water Marigold.
Is in Ipswich River.
Has been collected by G. D. Phippen.

Maruta Cotula. DC. — Mayweed.
Common. Yards and roadsides.

---

* The energy with which this plant establishes itself has often appeared remarkable. when I have found the vigorous plants growing in the crevices of the bark of trees, three or four feet above the ground, where the seeds have been deposited by the water, when the pond by which they stood was unusually full. A persevering root had in every case followed the retiring water, till it had finally reached the earth.

Yarrow.	**Achillea millefolium.** L.	

Common. Always well known.

Sneezewort, Goose-tongue.	**A. ptarmica.** L.	

Occasional. Generally tenants the vicinity of streams, but sometimes dry spots, as on the ridge of Pine Hill, by the Dungeon Road. Neither Gray or Bigelow allude to its slender habit, with us it has little more strength than a Galium.

Ox-eye Daisy, White-weed.	**Leucanthemum vulgare.** Lam.	

Common. Universally known and detested by cultivators.

Tansy.	**Tanacetum vulgare.** L.	

Common. Escaped from gardens and getting troublesome.

Shore Mugwort	**Artemisia caudata.** Mx. (?)	

I have found sparingly on the beaches at Nahant.

Mugwort.	**A. vulgaris.** L.	

Very rare. A few plants grew to great perfection in the garden of Jesse Rhodes, Esq., in 1855. I have never found it elsewhere.

Winged Everlasting.	**Gnaphalium decurrens.** Ives.	

Rare. I have met a few specimens in cartways in the woods; not more than three or four in all.

Sweet-scented Everlasting.	**G. polycephalum.** Michx.	

Abundant. Hills and pastures in light soils.

Low Cudweed.	**G. uliginosum.** L.	

Abundant. Cartways and neglected walks, in the ruts and broken sward.

Pearly Everlasting.	**Antennaria margaritacea.** R. Br.	

Frequent. Uplands and bushy fields.

Mouse-ear, Cat's Paw, Plantain-leaved Everlasting.	**A. plantaginifolia.** Hook.	

Common. Almost everywhere, unless among strong grass. Flowers earlier than anything else, not excepting *Saxafraga Virginiensis*.

Fire-weed.	**Erechthites hieracifolia.** Raf.	

Abundant. Certain to appear whenever the soil is stirred or burned in the woods.

Senecio vulgaris. L. — Common Groundsel.
Common. A vexatious weed in gardens, and by no means scarce in other places.

S. aureus. L. — Golden Ragwort, Golden Senecio.
Frequent. Wet places by streams. Grows profusely at the head of the Bowler Swamp, opposite Flax Pond.

Centaurea nigra. (?) L. — Knapweed.
Rare. I found one specimen on the railroad in Swampscott, in 1848; but no more. It grows abundantly on the turnpike at North Chelsea.

Cirsium lanceolatum. Scop. — Common Thistle.
Common. Roadsides and fields.

C. discolor. Spreng. — Two-colored Thistle.
Rare. Only to be found to my knowledge at the west end of Summer Street.

C. pumilum. Spreng. — Pasture Thistle.
Common. Woods and fields, widely distributed.

C. arvense. Scop. — Canada Thistle.
Frequent. Generally by roadsides, not specially odious here.

Onopordon acanthium. L. — Cotton Thistle.
Frequent. It seems partial to beds of decayed sea manure; it flourishes at Little Nahant, and generally where such manure has lain.

Lappa major. — Burdock.
Common. Too familiar to need further notice.

Lapsana communis. L. — Nipplewort.
A weed in gardens in S. Salem.—*Russell.*

Cichorium intybus. L. — Wild Succory, Chicory.
Abundant. Roadsides and railroad banks.

Krigia Virginica. Willd. * — Dwarf Dandelion.
Frequent. Only in the hills, where it starts in every open spot, if the ground be broken.

---

* Noticeable for the tint of the flowers, which are a full orange. No other plant exhibits it here, I believe, unless it is the Celandine, *Chelidonium majus.*

False Dandelion, Horse Dandelion.
Leontodon autumnale. L.
Common. Almost universally present in this section. I know no other plant, but the grasses, which is so fully distributed.

Canada Hawkweed.
Hieracium Canadense. Michx.
Occasional. Found at Linwood.

Rough Hawkweed.
H. scabrum. Michx.
Abundant. One of the familiar things of the woods and fields.

Rattlesnake weed.
H. venosum.
Frequent. Sometimes in woods, but more frequently in open pastures. The veined leaves very attractive.

Rattlesnake root.
Nabalus albus. Hook.
Frequent. It appears in many shady copses in stony slopes.

Tall white Lettuce.
N. altissimus. Hook.
Abundant. Shady woods, especially if damp and free from pines.

Dandelion.
Taraxicum Dens-leonis. Desf.
Common. I think this plant flowers longer than any we have.

Wild Lettuce.
Lactuca elongata. Muhl.
Common. Pastures and woods; the *Var. Sanguinia* being most frequent.

Mulgedium.
Mulgedium leucophæum. DC.
I found on the beaches at Gloucester harbor, Sept. 14, 1866.

Common Sow Thistle.
Sonchus oleraceus. L.
Rare. Now and then appears in old cultivated grounds, as along Boston Street.

Prickly Sow Thistle.
S. asper. Vill.
Rare. A few specimens have been secured about Berry's Mill, at Waterhill.

LOBELIDS.

(LOBELIA FAMILY.)

Lobelia cardinalis. L. — Cardinal Flower.
Frequent. Swampy localities and banks of streams.

L. inflata. L.* — Lobelia, Indian Tobacco.
Frequent. Wood roads and neglected fields.

L. spicata. Lam. — Pale spiked Lobelia.
Frequent. Meadow grounds and sometimes in upland fields.

L. Dortmanna. L. — Water Lobelia.
Rare. I have found it in Spring Pond and in Essex Pond, Hamilton; also in Middleton Pond. It usually grows in about two feet of water.

CAMPANULIDS.

(HAREBELL FAMILY.)

Campanula aparinoides. Pursh. — Slender Bell-flower.
Occasional I once found it in a meadow near Breed's Mills, but nowhere else in Lynn that I remember. It grows in the meadows in Lynnfield.

C. ROTUNDIFOLIA, L., the Harebell, appears to have been collected near Sutton's Mills, N. Andover, in 1850.—*Proc. Ess. Inst.*, 1856.

C. GLOMERATA, L., apparently overlooked by both Gray and Bigelow, is said to have been long established in Dark Lane, Danvers.—*Ibid.*
It is also abundant along the Newburyport Turnpike in the south part of Topsfield.

Specularia perfoliata. DC. — Clasping Specularia.
Frequent. At Three Needle Rock, near Saugus River; also through the woods in thin soil upon ledges.

---

* No more a virulent poison than a mustard plant. I have eaten it, drank it, slept upon it, rubbed and bathed with it, and never saw the slightest evil from its use; had it been poisonous, I must have died long since, unless I possess the stomach of a Mithridates.

KALMIDS.

(HEATH FAMILY.)

Dangleberry.
Gaylussacia frondosa. Torr & Gr.
Occasional. In Pratt's Pasture. near the pond and below a precipitous ledge. Also scattered through the woods, near the various swamps.

Huckleberry.
G. resinosa. Torr & Gr.
Abundant. Covering acres of ground. and very fruitful when young.

Cranberry.
Vaccinium macrocarpon. Ait.
Abundant. Meadows and ponds. In one place at Breed's Mills it formerly grew on a dry upland.

V. VITIS-IDÆA. L., the Cowberry has one or two localities in Danvers.—*Proc. Ess. Inst.*. 1856.

Low Blueberry.
V. Pennsylvanicum. Lam.
Abundant. Our first berry and one of the most fruitful. though not the finest.

Blue Huckleberry.
V. vacillans. Sol. (?)
Occasional. Very common on the ridge of Blood's Swamp Hills, and other like situations.

Highbush Blueberry.
V. corymbosum. L.
Abundant. Borders of swamps, forming thickets. and bearing superb fruit. Flourishing perfectly in Salem Pastures.

Black Blueberry.
V. fuscatum. Ait.
Abundant. Intermixed with *V. corymbosum*, and only distinguishable by its flowers or fruit.

Mountain Cranberry.
Arctostaphylos Uva-ursi. Spreng.
Occasional. A fine patch grows in Pratt's Pasture. and I have it from Rail Hill. near the cemetery. The finest growth I know of is on the ledges at Swain's Pond. Melrose.

Trailing Arbutus, Mayflower
Epigæa repens. L.
Very rare. It has been found in the woods between Lynn and Lynnfield.

Galutheria procumbens. L.*
Partridge-berry. Checkerberry.

Abundant. A large spot west of Pan Swamp Meadow is covered with it; but fruit is nowhere plenty.

Andromeda ligustrina. Muhl.
Privet Andromeda.

Abundant. One of our commonest water-shrubs, almost exactly like a blueberry, except in flowers and fruit.

A. POLIFOLIA, L., the Rosemary Andromeda, has a famous locality at Cedar Pond, Wenham. †

A. calyculata. L.
Rusty-leaved Andromeda.

Occasional. The best locality perhaps, is at the head of Bowler Swamp, between Boston and Chestnut Streets.

Clethra alnifolia. L.
Sweet Pepper-bush, White Alder.

Abundant. Another of the most common tenants of the swamp, and one of the most beautiful.

Rhodora Canadensis. L.
Rhodora.

Rare. Not in Lynn to my knowledge, except in one spot on the border of Breed's Pond. A choice plant for cultivation.

Azalea viscosa. L.
Swamp Pink, White Honeysuckle.

Abundant. Borders of ponds and damp situations.

---

* I desire in all cases to give the common or English name which is the best and most widely known; bnt I cannot engage that *Partridge-berry*, *Checkerberry*, or *Wintergreen*, mean anything definite, or apply to one plant more than another. These, like many others, are differently used by different persons. The student should know this and regulate his confidence in them accordingly. As to the present case, the *Gaultheria* is the plant used for flavoring confectionery and essences.

† This is one of the shrubs so pleasant to buy of the metropolitan florists. As much as three dollars has been paid for a single bush, when for half that sum a decent wagon-load could be brought from Wenham. It makes one think of Pindar's cheeses:—

" Where they were made, they sold for the immense
 Price of three sous apiece;
 But as salt water made their charms increase,
 In England the fixed rate was eighteen pence."

Mountain Laurel.	**Kalmia latifolia.**
Very rare. I have seen fresh specimens gathered in the vicinity of Pine Hill, but where, I could never ascertain. Very fine at Beverly, and magnificent at Gloucester.
Sheep Laurel, Lambkill.	**K. angustifolia.** L.
Common. Watercourses and damp lands everywhere.
K. GLAUCA, the Pale Laurel, is at Cedar Pond, Wenham.—*S. P. Fowler. Proc. Ess. Inst.*, 1856.
Canker Lettuce. Round-leaved Pyrola.	**Pyrola rotundifolia.** L.
Common. Shady places and woods, everywhere unless the soil be extremely poor.
Broad-leaved Pyrola.	**P. asarifolia.** Michx.
Occasional. Cold, shady woods. I have found it most in Pirates' Glen.
(I am not confident as to this species, as it runs, so near *P. chlorantha*, in some of its forms.)
Thin-leaved Pyrola.	**P. elliptica.** Nutt.
Frequent. Bears more exposure than *P. asarifolia*, but generally thrives only under thick pines.
Small Pyrola.	**P. chlorantha.** Swartz.
Rare. Pirates' Glen and a few other shady spots.
One-sided Pyrola.	**P. secunda.** L.
Rare. This, like the last, I have found in Pirates' Glen, but seldom elsewhere in this vicinity.
One-flowered Pyrola.	**Moneses uniflora.** Salisb.
Very rare. My only specimens were from Saugus, but it is now extinct there, and has since been found near Dungeon Rock. Quite frequent in the James Newhall pasture, between E. Saugus and the Centre.
Prince's Pine.	**Chimaphila umbellata.** Nutt.
Abundant. Shady, cool woods in good soil.
Spotted Wintergreen.	**C. maculata.** Pursh.
Is sparingly found on the south side of Middleton Pond. Also in Saugus.—*Dr. Clark.*

Hypopitys lanuginosa. Nutt. — Pine-sap, False Beech-drops.
Occasional. To be found in pine woods, bursting up through decaying leaves.

Monotropa uniflora. L. — Indian Pipe.
Frequent. Generally in pine woods, in the deep loose mould.
Var. MORISONIANA, Mx., with erect flowers, is frequent, appearing at the close of the season.

AQUIFOLIDS.
(HOLLY FAMILY.)

Prinos verticillatus. L. — Black Alder.
Abundant. Swamps and pond borders.

Ilex glabra. Gray. — Inkberry.
I have found abundant in the Magnolia Swamp, Gloucester.

Nemopanthes Canadensis. — Mountain Holly.
Occasional. Cool places in rich, alluvial soil.

PLANTIDS.
(PLANTAIN FAMILY.)

Plantago major. L. — Plantain.
Common. Familiar to all.

P. lanceolata. L. — Narrow Plantain.
Common. Gardens and fields.

P. maritima. L. — Seaside Plantain.
Frequent. Marshes. Rocks washed by the sea along the whole coast. In seeking for this, the student is liable to be misled, as I was, formerly, by the spikes of the Marsh Arrow-Grass, which see further on.

PLUMBIDS.
(LEADWORT FAMILY.)

Statice Limonium. L. — Marsh Rosemary.
Frequent. On the salt marshes within the range of high tides.

PRIMULIDS.

(Primrose Family.)

Star-Flower. **Trientalis Americana.** Pursh.
Abundant. I know no better locality than the summit of Second Pine Hill.

Upright Loosestrife. **Lysimachia stricta.** Ait.
Common. Damp grounds and brooksides. The bulbiferous state is not unusual.

Four-leaved Loosestrife. **L. quadrifolia.** L.
Common. Inhabits meadows and hilltops indiscriminately.

Lance-leaved Loosestrife. **L. lanceolata.** Walt.
Rare. Formerly abundant in a low place between Carnes and Federal Streets. The only locality I have known.

Tufted Loosestrife **Naumbargia thyrsiflora.** Reich.
I have found in the swampy woods by the Turnpike. east of Spring Pond.

Pimpernel **Anagallis arvensis.** L.*
Frequent. Cultivated grounds and dry fields. Also on the rocks at Nabant. fine specimens.

Inflated Featherfoil. **Hottonia inflata.** Ell.
Frequent. Swampy edges of Breed's Pond and the tributary brooks.

UTRICULIDS.

(Bladderwort Family.)

Inflated Bladderwort. **Utricularia inflata.** Walt.
Frequent. Lily Pond and others of its class.

Purple Bladderwort. **U. purpurea.** Walt.
Rare. Hardly in Lynn. A stagnant pond east of Lynnfield Hotel furnishes it in abundance. Ipswich River.—*Phippen.*

---

* Perhaps there is no native flower whose color comes more nearly to a pure red than this. The remarkable range of hues attributed to it by Gray seems to have no example so far as this region is concerned.

U. vulgaris. L. (?) — Common Bladderwort.
Common. In all ponds and sluggish waters.

U. intermedia. Hayne. — Creeping Bladderwort.
Occasional. Muddy sloughs near turnpike at Glenmere; also plentiful in a meadow reclaimed from Bowler Swamp, near the Cemetery.

U. CARNUTA, Michx., the Horned Bladderwort, a singular, leafless species grows scantily in springy places at Norman's Woe, Gloucester.
Also at Rocks Pasture, Lynn, and in Lynnfield. —*Chute.*

OROBANCHIDS.
(BROOMRAPE FAMILY.)

EPIPHEGUS VIRGINIANA, Bart., Beech Drops or Cancer Root, is very abundant in the beech woods of Hamilton, southwest of Essex Pond.

Aphyllon uniflorum. Torr. & Gr. — One-flowered Cancer-root.
Very rare. I once found a single specimen near Breed's Mills, and a few more on Burrill's Hill.

LINARIDS.
(SNAPDRAGON FAMILY.)

Verbascum Thapsus. L. — Mullein.
Common. Perfectly known.

Linaria Canadensis. Spreng. — Canada Snapdragon.
Common. Roadsides and dry fields.

L. vulgaris. Mill. — Toad Flax, Butter and Eggs.
Common. Roadsides and like places.

L. ELATINE, Miller. The Halbert-leaved Toad-flax is naturalized at Ipswich.—*Proc.Ess.Inst.*,1856.

Scrophularia nodosa. L. — Figwort.
Very rare. I doubt if in Lynn at all. It was formerly in one spot on Washington Street. I have found it at North Andover.

Chelone glabra. — Balmony, Snake-head.
Low ground and open swamps.

Monkey-flower. **Mimulus ringens.** L.
Frequent. Along streams; especially Strawberry Brook, below Bowler Meadow.

Hedge-Hyssop. **Gratiola Virginiana.** L.
Frequent. Wet places, pond borders, associated with the next.

Golden Hedge-Hyssop. **G. aurea.** Muhl.
Frequent. Generally in the same places as the previous, and hard to distinguish except by the color of the flower.

False Pimpernel. **Ilysanthes gratioloides.** Benth.
Occasional. Muddy borders of ponds and streams.

Water Speedwell. **Veronica Anagallis.** L.
Abundant. So thick in many ditches as to almost choke them up. Margins of Breed's Pond and like places.

Marsh Speedwell. **V. scutellata.** L.
Frequent. Wet places. Less plentiful than *V. Anagallis.*, and apparently much later.

Common Speedwell. **V. officinalis.** L.
I found, in 1876, a fine patch by the roadside at "Burrill's Landing," S.W. of Dungeon Rock.

Thyme-leaved Speedwell. **V. serpyllifolia.** L.
Common. Creeps among the grass in cool ground.

Neckweed. **V. perigrina.** L.
Abundant. A troublesome weed in gardens.

Corn Speedwell. **V. arvensis.** L. (?)
Rare. A few depauperate plants grew formerly by Holmes' Mill, Walnut Street. On gravelly banks at Nahant.

Purple Gerardia. **Gerardia purpurea.** L
Abundant. Meadows and damp lands.
G. MARITIMA, Raf. The Seaside Gerardia. I found in the autumn of 1861, a fine patch on the marsh between Sagamore Hill and Beach Street, near Long Beach. In good flower and well determined.

G. tenuifolia. Vahl.   Slender Gerardia.
Abundant. Uplands and warm pasture slopes.

G. flava. L.   Downy Gerardia.
Occasional. In the E. section, near Swampscott.

G. pedicularis. L.   Bushy Gerardia
Frequent. Formerly near Echo Grove.

Castilleia coccinnea. Spreng.   Painted Cup.
The yellow variety of this abounds in a meadow near the Meeting House, Topsfield.

Pedicularis Canadensis. L.   Lousewort, Wood Betony.
Frequent. Sunny banks among bushes, generally in rich, moist soil.

P. lanceolata. Mx.   Yellow Lousewort.
I think I found this near Shute's Brook, above the railroad station, Saugus Centre.

Melampyrum pratense. L.   Cow Wheat.
Var. AMERICANUM. Benth.
Abundant. Shady woods and pastures. A difficult thing to transplant, withering even when taken up with a large ball of earth.

VERBENIDS.

(VERVAIN FAMILY)

Verbena hastata. L.   Blue Vervain.
Very rare. I have not seen it in Lynn. Reported by *Dr. Holder* in the neighborhood of Chatham Street, above Essex. Also at Saugus.

V. urticifolia. L.   Nettle-leaved Vervain.
Common. Roadsides and fields.

Phryma leptostachya. L.   Lopseed.
Is found on the Newburyport Turnpike, near Middle Road, Danvers.

MENTHIDS.

(MINT FAMILY.)

Mentha viridis. L.   Spearmint.
May be found at Stocker's Brickyard, E. Saugus.

Peppermint.	**M. piperita.** L. Abounds at the Foundry on Weare River, Hingham.
Meadow-mint.	**M. Canadensis.** L. Abundant. Meadows and brooksides.
Water Hoarhound.	**Lycopus sinuatus.** Ell. Abundant. Wet grounds and streams.
Pennyroyal.	**Hedeoma pulegioides.** Pers. Frequent. Neglected fields to some extent, but more in pastures and open woods. Smaller than specimens from further inland.
Wild Bergamot.	**Monarda fistulosa.** L. Grows at N. Andover on the southwest side of Great Pond.
Catnip.	**Nepeta cataria.** L. Rare. Sparingly at Nahant, and occasionally in Lynn, but always near some old garden.
Ground Ivy, Gill-over-ground.	**N. Glechoma.** Benth. Frequent. About dwellings and old garden walls.
Wild Basil.	**Pycnanthemum incanum.** Michx. Rare. Only found thus far on the southern slope of Linwood.
Mountain Mint.	**P. muticum.** Pers. Very rare. Grows sparingly in damp spots in Swampscott. THYMUS SERPYLLUM, L., the Garden Thyme, grows in Boxford.—*Proc. Ess. Inst.*, 1856.
Self-heal.	**Prunella vulgaris.** L. Common. Generally an upland plant, but very easily suited with soil.
Scull-cap.	**Scutellaria galericulata.** L. Rare. In the eastern part of the city.—*True.* Grows finely at Lily Pond, Manchester. Also at Peters' Meadow, Danvers.

S. lateriflora. L.   Side-flowering Scullcap.
Frequent. Gravelly banks of ponds.

Lamium purpureum. (?) L.   Henbit, Dead Nettle.
Occasional. An intrusive weed in old gardens as those on Boston Street, near Federal.

Leonurus cordiaca. L.   Motherwort.
Common. Familiar to all.

Galeopsis tetrahit. L.   Common Hemp-Nettle.
Rare. Grew formerly between Federal and Mall Streets.
STACHYS PALUSTRIS, L., the Marsh Hedge Nettle, appears in Ipswich and elsewhere.—*Proc. Ess. Inst.*, 1856.

Trichostema dichotomum. L.   Blue Curls, False Penny-royal.
Common. Pastures and fields, especially in the hills.
A variety with red flowers occurs on the railroad between S. Reading and Lynnfield Hotel.—*A. P. Chute.*

Teucrium Canadense. L.   Germander, Wood Sage.
Occasional. I find it only on upland hillocks and margins, about the marshes east of the Long Railroad Bridge, Saugus River. I found, in 1859, a form with pink flowers growing very rankly round old yards in N. Andover.

SYMPHYTIDS.

(BORAGE FAMILY.)

Echium vulgare. L.   Blue-weed.
Very rare. A few specimens found on Walnut Street on a manure bed.
Seed no doubt brought from abroad.—*Dr. Nye.*

Myosotis laxa. Lehm.   Forget-me-not.
Occasional. Pirates' Glen; other wet situations.

M. stricta. Link.   Scorpion-grass.
Abundant. A little weed on the hillsides, of small consequence.

Beggar's Lice. **Cynoglossum Morisoni.** DC.
Found on the south side of Flax Pond, in one instance.

Stickseed. **Echinospernum Lappula.** Lehm.
Is at N. Andover, 1859.—*Russell.*

CONVOLVULIDS.

(MORNING-GLORY FAMILY.)

Hedge Bindweed. **Calystegia sepium.** R. Br.
Frequent. Climbing on walls in many places.

Bindweed. **Convolvulus arvensis.** L.
Frequent. Formerly abundant near Ingalls' Pond, Fayette Street. Also in cultivated land about N. Federal Street.

CUSCUTA EPILINUM, Weihe., the Flax Dodder, has been found in Rowley.—*Proc. Ess. Inst.*, 1856.

Common Dodder. **Cuscuta Gronovii.** Willd.
Frequent. Damp and weedy situations, climbing on Bidens, Polygonums, etc.

SOLANIDS.

(NIGHTSHADE FAMILY.)

Thorn-apple, Jamestown-weed. **Datura Stramonium.** L.
Common. Old grounds and rich places. It seems partial to beds of decayed sea-manure, and is commonest near the beaches.

Black Henbane. **Hyoscyamus niger.** L.
Very rare. At the north end of Short Beach, Nahant, is the principal locality. A plant now and then in old, rich yards.

Apple of Peru. **Nicandra physaloides.** Gært.
Rare. I have seen it in yards in Summer and S. Common Streets. but sparingly. Becoming more plentiful as an intruder in gardens in West Lynn.

Solanum dulcamara. L.*   Woody Nightshade.
Abundant. Brooksides and thickets.

S. nigrum. L.   Common Nightshade.
A few plants have been found on Short Beach, Nahant, and also on Lynn Common.

GENTIANIDS.
(GENTIAN FAMILY.)

Gentiana crinita. Frœl.   Fringed Gentian.
Occasional. Fresh marsh and like situations.

G. Andrewsii. Grise.   Closed Gentian.
Very rare. Fay estate. Have considerable doubt as to this; it seems not to be the common species of N. England, known as *G. saponaria.*

Menyanthes trifoliata. L. †   Buck Bean.
Rare. Plentiful along Shute's Brook at the railroad station, Saugus Centre. Also in a muddy slough in the eastern part of Rocks Pasture.

Limnanthemum lacunosum. Grise.   Floating Heart.
Frequent. Ponds, with the different water lilies.

APOCYNIDS.
(DOGSBANE FAMILY.)

Apocynum androsæmifolium. L.   Dogsbane, Flytrap.
Common. Woods and uplands.

---

* This plant is often spoken of as the Deadly Nightshade, a name which properly belongs to the *Atropa Belladonna,* L. The two are readily distinguished by the fruit, which in the *Atropa* is black, but a brilliant red in the other. The true Deadly Nightshade is a rare plant in all parts, and the *Solanum Nigrum,* L., the Common Nightshade, which somewhat resembles it, is not, I think, to be found in this vicinity.

† Any one who has access to a brookside or other wet place may cultivate this elegant plant without trouble; and be rewarded with flowers as fine as hyacinths.

ASCLEPIDS.

(Milkweed Family.)

Common Milkweed.
Asclepias Cornuti. Decaisne.
Common. Fencerows and edges of cultivated lands.

Poke Milkweed.
A. phytolaccoides. Pursh.
Occasional. Sheltered pastures and clearings in the woods.

Four-leaved Milkweed.
A. quadrifolia. Jacq. *
Rare. I have met with it once or twice in Dungeon Pasture, and the neighboring lands.

Swamp Milkweed, Indian Hemp.
A. incarnata.
Common. Watercourses and low grounds: sometimes in uplands, but more rarely.

Whorled Milkweed.
A. verticillata. L.
Was found in 1861 at the entrance of Pine Grove Cemetery, by Mr. Edward Johnson. Only known to me before at Waitt's Mount, Malden. I found it in 1862, on the rocky hill at Edward Stone's house, Boston Street, near the Cemetery in considerable quantity, with *Sempervivum*.

OLIDS.

(Olive Family.)

Privet, Prim.
Ligustrum vulgare. L.
Common. Everywhere on the hills in the southern part of the town.

White Ash.
Fraxinus Americana. L. (?)
Occasional. Hardly as frequent as the next, though I confess to some indecision about both species.

Black Ash.
F. sambucifolia. Lam. (?)
Woods and borders of fields, as at Linwood.

---

* I do not understand how this plant has so long escaped cultivation. Its beauty cannot be questioned, and it would no doubt endure a garden soil perfectly.

APETALOUS EXOGENS.

CHENOPIDS.
(Goosefoot Family.)

Salsola kali. L. — Saltwort.
Common. On the various beaches near high-water mark.

Suæda maritima. Moquin. — Salt Goosefoot.
Abundant. Borders of all the salt marshes.

Salicornia herbacea. L. — Samphire.
Common. On the marshes wherever the sea has access.
SALICORNIA AMBIGUA, Michx., the Creeping Glasswort, may be found at Gloucester.—*Proc. Ess. Inst.*

Chenopodium glaucum. L. — Oak-leaved Goosefoot.
I found in 1861, in the barnyard of Isaiah Hill, Summer Street. Also previously at Marblehead Neck, near the beach.

C. hybridum. L. — Maple-leaved Goosefoot.
I found at North Andover in 1859.

C. album. L. — Pigweed.
Common. A weed in every garden.

C. urbicum. L. — Triangular-leaved Goosefoot.
Occasional. About dwellings and yards.

Ambrina anthelmintica. Spach. — Wormseed.
Rare. Found on a manure bed on Holyoke Street. Also in Glenmere.

Atriplex patula. L. — Spreading Orache.
Abundant. Along the Eastern Railroad, and the margin of all marshes.
A plant of this family, which seems undescribed by Gray, grows on Short Beach, Nahant. It may possibly be an *Obione*.

AMARANTHIDS.

(AMARANTH FAMILY.)

Too ugly to have a name.
Amaranthus albus. L.
Common. A pertinaceous intruder into every cultivated spot.

Mongrel Amaranth.
A. hybridus. L.
Common. As homely as its fellow, and much more troublesome.

PHYTOLIDS.

(POKEWEED FAMILY.)

Poke, Scoke, Gargel.
Phytolacca decandra. L.
Frequent. Wood clearings, especially where fires have been.

POLYGONIDS.

(SMARTWEED FAMILY.)

Tall Persicaria.
Polygonum Pennsylvanicum. L.
Occasional. Pine Grove Cemetery has furnished fine plants nearly four feet high.

Lady's Thumb. Heart's Ease.
P. persicaria. L.
Common. Manure beds and rich waste grounds. Easily distinguished by the dark spot on the leaf.

Smartweed.
P. hydropiper. L.
Common. Cool and moist situations. Well known.

Mild Water-Pepper.
P. hydropiperoides. Michx.
Frequent. Swamps and brooks.

Water Persicaria.
P. amphibium. L.
Occasional. Not as common as either of the last three. Have found the best specimens at Great Pond. N. Andover. Hardly think it is in Lynn.

Knotgrass.
P. aviculare. L.
Common. Inseparable from civilization. If any plant is developed from inorganic matter it must be this.

P. tenue. Michx. — Slender Knotgrass.
Rare. Now and then appearing in gravel pits, and along some by-roads.
POLYGONUM ARTICULATUM, L., *Var.* MULTIPLEX, the Jointweed has been met with at Ipswich and very abundantly at Groveland.—*Proc. Ess. Inst.* Also on the railroad between Lynnfield and S. Reading, abundantly.

P. arifolium. L. — Larger Scratchgrass.
Frequent. Dense and secluded thickets and swamps.

P. sagittatum. L. — Common Scratchgrass, Tear-thumb.
Common. Swamps and meadows.

P. convolvulus. L. — Black Bindweed. Wild Bean.
Common. Gardens and fields.

P. dumetorum. L. (?) — Climbing False Buckwheat.
Frequent. Running on walls and bushes.

Fagopyrum esculentum. Mœnch. — Buckwheat.
Occasional. Naturalized to some extent. I have found it on Lynn Common.

Rumex salicifolius. Hook. (?) — White Dock, Willow-leaved Dock.
Frequent. Nahant; on the beach and elsewhere.

R. obtusifolius. L. — Bitter Dock.
Common. Damp cultivated grounds.

R. crispus. L. — Curled Dock, Yellow Dock.
Common. Needs no description.

R. acetosella. L. — Common Sorrel.
Common. Pastures and fields. The abundance of the plant appears to be inversely as the richness of the soil.

LAURIDS.
(SASSAFRAS FAMILY.)

Sassafras officinale. Nees. — Sassafras.
Frequent. Seldom grows large with us. I have seen it ten or fifteen feet high at Baker's Hills, Saugus, and about the same size on a knoll in Pan Swamp Meadow.

Spice Wood, Fever-bush. **Benzoin odoriferum.** Nees.
Occasional. Pirates' Glen, and in the swamp along Stony Brook, near Linwood Street. The south side of Bennet's Swamp is a better locality still.

THYMELIDS.

(MEZEREON FAMILY.)

Mezereum. **Daphne Mezereum.** L.
Occasional. Paradise Woods, Swampscott.

SANTALIDS.

(SANDALWOOD FAMILY.)

False Toad-flax. **Comandra umbellata.** Nutt. *
Abundant. Rocky hills and pastures.

CALLITRIDS.

(WATER-STARWORT FAMILY.)

Water Starwort. **Callitriche verna.** L.
Frequent. Running waters with muddy bottoms.

Narrow-leaved Starwort. **C. linearis.** Pursh.
Stagnant ponds and slow streams.

EUPHORBIDS.

(SPURGE FAMILY.)

Sun Spurge. **Euphorbia helioscopia.** L.
A few plants found at E. Saugus in a lot on Boston Street, beside the railroad, south side of the street. Otherwise not found.

---

*The sharp-eyed student will notice the strange manner in which the anthers of this species are tied down with threads, and if he feels very wise, may try his ingenuity for the reason why it is so. There are mysteries in plants, and this is one of them.

E. Esula. L.   Leafy Spurge, Queen Anne.

Common. About old dwellings. An emigrant from the garden. Near Devil's Den, Newburyport, it flourishes in the gravel of the railroad in a very peculiar form.

E. polygonifolia. L.   Seaside Spurge.

Frequent. Beaches at Nahant, especially Long Beach.

Acalypha Virginica. L.   Three-seeded Mercury.

Common. Yards and rubbish heaps.

ULMIDS.

(ELM FAMILY.)

Ulmus Americana. L.   American Elm.

Frequent. Generally in wet places, but not always.

Celtis occidentalis. L.   Hackberry.

Is said to grow in Great Pastures, between Lynn and Salem.—*Robinson.*

Urtica dioica. L.   Common Nettle.

Occasional. Old gardens and weedy yards.

U. urens. L.   Small Nettle.

Occasional. Now and then to be found about old yards in the eastern part of the city.

Pilea pumila.   Clearweed, Stingless Nettle.

Common. Gardens and manure beds.

Boehmeria cylindrica. Willd.   False Nettle.

Occasional. Along the brook in Marion Street, and on the edge of several different streams.

Cannabis sativa. L.   Hemp.

Occasional. Old gardens and weedy yards.

PLATANIDS.
(Buttonwood Family.)

Buttonwood. **Platanus occidentalis.** L.
Frequent. Here and there in the woods, seemingly escaped from cultivation.

JUGLANIDS.
(Walnut Family.)

Pignut, Hickory. **Carya glabra.** Torr.
Abundant. Forms large tracts of young woods in Dungeon Pasture and vicinity.

QUERNIDS.
(Oak Family.)

White Oak. **Quercus alba.** L.
Common. Woods and copses.

Swamp Chestnut Oak. **Q. Primus.** L.
Formerly abundant in and about the "Johnson Swamp," between Laighton Street and Beacon Hill Avenue. Mostly destroyed.

Q. PRINOIDES, Willd., the Chinquapin Oak, was found on the Town Farm, Peabody, at a meeting of the Essex Institute, in June. 1857. A single bush was identified by me in 1886 on the summit of "Mount Tabor," near "Penny Brook," Lynn.

Scrub Oak. **Q. illicifolia.** Wang.
Frequent. Dry hills, forming rough thickets. alike destructive to the clothes and comfort of the wood ranger.

Swamp White Oak. **Q. bicolor.** Willd.
Common. Low grounds.

Quercitron Oak, Black Oak. **Q. tinctoria.** Bart.
Occasional. Woods in dry soil.

Scarlet Oak. **Q. coccinea.** Wang.
Frequent. A heavy tree, scattered about in open lands.

Castanea vesca. L.   Chestnut.
    Rare. Grows finely in Lynnfield, but not at all, as is believed, in Lynn.

Fagus ferruginea. Ait.   Beech.
    Occasional. Scattered through the Ox Pasture, toward Lynnfield. A tree or two grow in Pratt's Pasture, and a few remain in Linwood.

Corylus Americana. Walt.   Hazel Nut.
    Frequent. Old pastures and fence rows. Fruit seldom plentiful.

C. rostrata. Ait.   Beaked Hazel Nut.
    Grows in Saugus and has been produced at Field Meetings of the Essex Institute.

Carpinus Americana. Mx.   American Hornbeam.
    Rare. Formerly common in Linnean Grove, but hardly anywhere else.

Ostrya Virginica. Willd.   Lever Wood, Hop Hornbeam.

MYRIDS.

(BAYBERRY FAMILY.)

Myrica Gale. L.   Sweet Gale, Meadow Fern, Dutch Myrtle.
    Abundant. Apparently confined to the western end of the township.

M. cerifera. L.   Bayberry.
    Common. Pastures and woodlands.

Comptonia asplenifolia. Ait.   Sweet Fern.
    Common. Perfectly familiar.

BETULIDS.

(BIRCH FAMILY.)

Betula populifolia. Ait.   White Birch.
    Common. Formerly covering many acres near the house of N. C. Hutchinson.

B. papyracea. Ait.   Paper Birch, Canoe Birch.
    Is abundant, as I think along the Spicket River, in Methuen.

Red Birch, River Birch.	**B. nigra.** L. (?)
Occasional. Three trees stand on the side of Forest Rock, in Pine Grove Cemetery, which do not appear to be *B. excelsa,* and certainly are not *B. lenta.* If they are not this species, they can hardly be included in Gray's enumeration.
Yellow Birch.	**B. excelsa.** Ait.
Frequent. Scattered through the woods and swamps.
Cherry Birch, Black Birch.	**B. lenta.** L.
Rare. I have never found a specimen in Lynn, but it has been gathered in former times by others, and may still be found sparingly.
A low birch grows on the shores of Great Pond, N. Andover, which is either *B. pumila,* L., or else the dwarf *B. papyracea,* which, *quere.*
Speckled Alder.	**Alnus incana.** Willd.
Is plentiful along Shute's Brook, near the railroad station, Saugus Centre. And likewise in the swamps below the residence of E. B. Phillips, Swampscott.
Common Alder.	**A. serrulata.** Ait.
Common. Sure to appear wherever there is water.

POPULIDS.

(WILLOW FAMILY.)

Low Bush Willow.	**Salix humilis.** Marshall.
Frequent. I suppose this, and possibly *S. tristis,* Ait., and *S. discolor,* Muhl, are so blended in our flora as to make good distinction difficult.
Swamp Willow, Pussy Willow.	**S. eriocephala.** Michx. (?)
Frequent. Meadows and streams, sometimes migrating to the uplands. The "Pussy Willow" may comprise more than one species, and I am not well satisfied as to this one in particular.
Hoary Willow.	**S. candida.** Willd.
A small bush near Holmes' Pond, and another on the hillside east of Breed's Pond, are of this species possibly. They are quite distinct, and I cannot otherwise determine them.

S. alba. L.   Yellow Willow.
Var. VITELLINA.
Common. Thoroughly naturalized in wet situations. Probably several varieties are found with us.

Populus tremuloides. Michx.   American Aspen.
Frequent. Scattered through the woods, but seldom attaining any great size.

P. grandidentata. Michx.   Tooth-leaved Poplar.
Frequent. Woods near Breed's Pond. The most elegant poplar we have.

GYMNOSPERMOUS EXOGENS.

PINIDS.

(PINE FAMILY.)

Pinus rigida. Miller.   Pitch Pine.
Common. Extensively distributed all over the dry and rocky pastures in the southern part of the territory.

P. strobus. L.   White Pine.
Common. Not partial to light soils, but growing vigorously in low lands. Greatest height with us about one hundred feet.

Abies Canadensis. Michx.   Hemlock.
Frequent. Scattered, or forming groves, but rarely attaining much magnitude.

Larix Americana. Michx.   Black Larch, Hacmatac.
Occasional. Grows in Tomlin's Swamp and similar places toward Lynnfield. Seldom large.

White Cedar. **Cupressus thyoides.** L.
Frequent. Fills a swamp at Swampscott, and another, of fine trees, near Birch Brook Pond, near Mr. H. B. Newhall's farm-house.

Juniper,
Horse Savin. **Juniperus communis.** L.*
Common. Pastures and hills.

Red Cedar. **J. Virginiana.** L. †
Common. The undisputed tenant of all rocky and sterile hills.

ENDOGENS.

ARIDS.

(Wake-Robin Family.)

Indian Turnip. **Arum triphillum.** L.
Adundant. In every wet, rich thicket. The striped or variegated form is frequent here, as in other places.

Arrow Arum. **Peltandra Virginica.** Raf.
Occasional. Along the banks of brooks.

Water Arum. **Calla palustris.** L. ‡
Frequent. At Penny Bridge Brook, in large quantity.

---

* The largest specimen I have seen formerly grew on Second Pine Hill, being some two hundred feet in circumference. It was burnt in 1852.

† A tree stood formerly on what is known as Pigsty Rock or Big Cedar Hill, worthy to be styled the Patriarch. It measured over seven feet in girth in 1844, with a proportionate height.

‡ I am led to suspect that many are unware that the secluded waters afford this younger sister of the fine *Calla Ethiopica* of the greenhouse. It is somewhat more modest but hardly less charming than the other, and would prove an ornament to the aquarium or artificial pond.

Symplocarpus fœtidus. Salisb. — Skunk Cabbage
Abundant. In all wet grounds. Familiar.

Acorus calamus. L. — Sweet Flag.
Occasional. Most easily found on the brackish banks of Strawberry Brook, just above Cottage Street.

TYPHIDS.
(CAT-TAIL FAMILY.)

Typha latifolia. L. — Cat Tail.
Frequent. Fresh marshes and stagnant ponds.

Sparganium ramosum. Hudson. — Burr Reed.
Frequent. Ditches and muddy streams. A ditch in Bridge Street gives good specimens.

S. Americanum. Nutt. — Smaller Burr Reed.
Abundant. In almost all muddy places.

S. natans. L. — Sparganium.
Frequent. Margins of Breed's Pond, and other like places.

ZOSTERIDS.
(SEA-WRACK FAMILY.)

Zostera marina. L. — Eel-grass.
Abundant. In the harbor, and every salt-water ditch beyond low-water mark.

Potamogeton natans. L. — Floating Pondweed.
Frequent. Most of the ponds.

P. heterophyllus. Schreber. — Slender Pondweed.
Frequent. Ponds in shallow water; Bartholomew's Pond, in Peabody, for instance.

P. lucens. L. — Thin-leaved Pondweed.
Occasional. Only known to be at Breed's Pond, but probably in other waters also. My specimens were all imperfect, so that the name is partly conjectural.

Clasping Pondweed.   **P. perfoliatus.** L.
Abundant. Strawberry Brook.

Narrow-leaved Pondweed.   **P. pauciflorus.** Pursh. (?)
Abundant. Strawberry Brook at the tan yard, with the last, forming dense mats on the water.

POTAMOGETON sp. (?) This is from Wenham Pond, and not a perfect specimen. For some time I regarded it as *P. perfoliatus*, but this idea was set aside by finding that species afterward. As it seems not to agree with any species noticed by Gray, I would give a description, but the specimens are unfortunately destroyed.

ALISMIDS.
(WATER-PLANTAIN FAMILY.)

Marsh Arrowgrass.   **Triglochin maritimum.** L.
Abundant. Scattered thickly over the salt marshes, forming annular patches. The spike might easily be taken for that of a plaintain.

Water Plaintain.   **Alisma plantago.**
Frequent. Water-courses and borders of ponds.

Arrowhead.   **Sagittaria variabilis.** Engelm.
Common. Everywhere in wet places. The innumerable forms are all included in this species.

ORCHIDS.
(ORCHIS FAMILY.)

Twayblade.   **Liparis Lœselii.** Rich.
Very rare. Formerly along Stacey's Brook.

Large Coral-root.   **Corallorhiza multiflora.** Nutt.
Rare. Scattered on Blood Swamp Hills, west of Dungeon Rock and so southward.

Late Coral-root.   **C. odontorhiza.** Nutt.
Rare. Rather plentiful on the east side of Edwards' Swamp.

Pale Green Orchis.   **Gymnadenia tridentata.** Lindl.
Rare. I have found a single specimen in Salem Pasture.

Platanthera flava. Gray. — Yellowish Orchis.
Occasional. In the neighborhood of Birch Pond.

P. blephariglottis. Lindl. — White Fringed Orchis.
Very rare. Not in Lynn. Possibly in the meadow lands at Swampscott, but this admits of doubt.—*Dr. Holder.*
It belongs to the flora of Cape Ann, rather than ours, being frequent in the Magnolia Swamp.

P. lacera. Gray. — Ragged Orchis.
Common. Our most common species, distributed largely in all damp lands.

P. psycodes. Gray. — Small Purple Fringed Orchis.
Occasional. Swamps and other wet situations.

P. fimbriata. Lindl. — Large Purple Fringed Orchis.
Very rare. Only one specimen reported.

P. orbiculata. Lindl. — Round-leaved Orchis.
Very rare. One specimen on the south side of Mt. Nebo, at "Pilgrim's Rest," Aug. 23, 1885.

Arethusa bulbosa. L. — Arethusa.
Rare. Confined to the eastern part of the city. I have it from Rock's Pasture.

Pogonia ophioglossoides. Nutt. — Adder's Tongue Arethusa.
Frequent. Meadows and bogs, widely distributed.

Calopogon pulchellus. R. Br. — Cymbidium.
Frequent. In company with the last, usually, but not quite as plentiful.

Spiranthes gracilis. Bigel. — Naked Spiranthes.
Frequent. Uplands among bushes. A plant of solitary habit.

S. cernua. Richard. — Ladies' Tresses.
Abundant. Cool meadows among the grass, which it much resembles in its leaves.

Goodyera repens. R. Br. — Pale Rattlesnake Plantain.
Frequent. Pine woods. Distinguished from the next by the white veins being duller, and the leaves more pointed.

Rattlesnake Plantain.	**G. pubescens.** R. Br. Frequent. Pine woods. Generally in patches.
Low Ladies' Slipper. American Valerian.	**Cypripedium acaule.** Ait. Abundant. Pine woods for the most part, but to be found on many of the hills.

AMARYLLIDS.

(AMARYLLIS FAMILY.)

Yellow Star-of-Bethlehem.	**Hypoxis erecta.** L. Abundant. Rich uplands and shady borders of woods.

IRIDS.

(BLUE-FLAG FAMILY.)

Large Blue Flag.	**Iris versicolor.** L. Abundant. In all wet lands. Well known.
Slender Blue Flag.	**I. Virginica.** L. Frequent. Wet meadows and borders of streams.
Blue-eyed Grass.	**Sisyrinchium Bermudiana.** L. Common. Rather partial to wet spots, but thrives anywhere in good soil.

SARSIDS.

(GREENBRIER FAMILY.)

Green Brier, Bull Brier.	**Smilax rotundifolia.** L. Common. Forms the most intricate of all thickets, and gets the hearty execration of all who have to pass through its meshes.
Carrion Flower.	**S. herbacea.** L. Frequent. Shady banks and meadow borders. The common name cannot be matched for expressiveness.
Nodding Trillium.	**Trillium cernuum.** L. Rare. Good plants along Shute's Brook, at railroad station, Saugus Centre.

T. erectum. L. — Birthroot.
 Also *Var.* AL.BUM, have been found by Jno. Sears in a swamp between Danvers and Wenham. The former common there.—*Russell.*

Medeola Virginica. L. — Cucumber Root.
 Frequent. Moist, shady thickets.

LILIDS.
(LILY FAMILY.)

Polygonatum pubescens. Pursh. — Small Solomon's Seal.
 Frequent. Warm, rocky slopes in light, rich soil.

Smilacina racemosa. Desf. — False Spikenard.
 Frequent. Stony woods in warm exposures.

S. stellata. Desf. — Star-flowered Solomon's Seal.
 Rare. Banks of Stacey's Brook, Swampscott.—*Dr. Holder.*
 It proves to be plentiful on a hillock at Willis' Neck.

S. bifolia. Ker. — Two-leaved Solomon's Seal.
 Common. Woods and bushy fields everywhere.
 CLINTONIA BOREALIS, the Northern Clintonia, is at Pleasant Pond, Wenham.—*Proc. Ess. Inst.*, 1856. And I found it at the Magnolia Swamp, Gloucester, in 1861.

Ornithogalum umbellatum. L. — White Star-of-Bethlehem.
 Rare. Occasionally found in Glenmere, in wet situations.

Allium Canadense. Kahn. — Wild Meadow Garlic.
 Occasional. Both the eastern and western parts of the city afford specimens.

Lilium Philadelphicum. L. — Red Lily.
 Frequent. Berry pastures among the bushes. Hard to find till it flowers.

L. Canadense. L.* — Yellow Lily.
 Occasional. Not as common as the last.

---

* A resplendent thing in cultivation. Several of my friends have had it attain a height of six feet, with a dozen or so of flowers open at once.

Dog-tooth Violet.
Erythronium Americanum. Smith.
   Occasional. Confined almost exclusively to the eastern section.

MELANTHIDS.

(INDIAN POKE FAMILY.)

Clasping Bellwort.
Uvularia perfoliata.
   Very rare. Formerly at Hathorne's Hill, but extirpated there. I find it even more thrifty in the Blueberry Pasture.

Common Bellwort.
U. sessilifolia. L.
   Abundant. Shady, open woods, often covering large areas.

White Hellebore, Indian Poke.
Veratrum viride. Ait.
   Occasional. Upper Swampscott. Neighborhood of Birch Pond.

JUNCALIDS.

(RUSH FAMILY.)

Bulrush.
Juncus effusus. L.
   Common. Bogs and meadows everywhere.

White-seeded Rush.
J. paradoxus. E. Meyer. (?)
   Abundant. Bogs and swampy meadows.

Black-grass.
J. Gerardi. Loisel.
   Abundant. Forms patches on the lightest part of the marshes.

PONTEDERIDS.

(PICKEREL-WEED FAMILY.)

Pickerel-weed.
Pontederia cordata. L.
   Abundant. Brooks and ponds, sometimes forming large clumps.

XYRIDS.

(YELLOW-EYED GRASS FAMILY.)

Xyris Caroliniana. Walt.   Yellow-eyed Grass.
Occasional. Rills in Rock's Pasture, and bogs along Stony Brook.

ERIOCAULIDS.

(PIPEWORT FAMILY.)

Eriocaulon septangulare. Wither.   Small Pipewort.
Frequent. Borders of ponds, growing in the edge of the water, or where it has recently dried away.

INDEX OF GENERA.

Abies	77	Aster	47
Abutilon	29	Atriplex	69
Acalypha	73	Azalia	57
Acer	32		
Achillea	52	Baptisia	35
Acorus	79	Barbarea	25
Actæa	23	Benzoin	72
Æthusa	42	Berberis	23
Agrimonia	36	Betula	75
Alisima	80	Bidens	51
Allium	83	Bochmeria	73
Alnus	76	Brasenia	23
Amaranthus	70		
Ambrina	69	Cakile	25
Ambrosia	50	Calla	78
Amelanchier	38	Callitriche	72
Ammannia	38	Calapogon	81
Ampelopsis	32	Caltha	22
Amphicarpæa	33	Calystegia	66
Anagallis	60	Campanula	55
Andromeda	57	Cannabis	73
Anemone	21	Capsella	25
Antennaria	52	Cardamine	25
Aphyllon	61	Carpinus	75
Apios	33	Carum	43
Apocynum	67	Carya	74
Aquilegia	23	Castanea	75
Aralia	43	Castillua	63
Archangelica	42	Ceanothus	32
Arctostaphylos	56	Celastrus	32
Arenaria	28	Celtis	73
Arethusa	81	Centauria	53
Artimisia	52	Cephalanthus	46
Arum	78	Cerastium	28
Asclepias	68	Chelidonium	24

Chelone	61	Euphorbia	72
Chenopodium	69	Eupatorium	46
Chimaphila	58		
Chrysosplenium	41	Fagopyrum	71
Cichorium	53	Fagus	75
Cicuta	42	Fragaria	37
Circæa	39	Fraxinus	68
Cirsium	53	Fumaria	24
Clematis	21		
Clethra	57	Galeopsis	65
Clintonia	83	Galium	45
Comandra	72	Gaultheria	57
Comium	43	Gaylussacia	56
Comptonia	75	Genista	35
Coptis	23	Gentiana	67
Convolvulus	66	Geranium	30
Corallorhiza	80	Gerardia	62
Coreopsis	51	Geum	36
Cornus	43	Gnaphalium	52
Corydalis	24	Goodyera	81
Corylus	75	Gratiola	62
Crantzia	42	Gymnadenia	80
Cratægus	38		
Cupressus	78	Hamamelis	41
Cuscuta	66	Hedeoma	64
Cynoglossum	66	Helianthemum	26
Cypripedium	82	Helianthus	51
		Hepatica	21
Datura	66	Heracleum	42
Daphne	72	Hieracium	54
Daucus	42	Honkenya	28
Desmodium	34	Hottonia	60
Dianthus	28	Houstonia	46
Diervilla	45	Hudsonia	26
Diplopappus	49	Hydrocotyle	41
Draba	25	Hyoscyamus	66
Drosera	27	Hypericum	27
		Hypopitys	59
Echenospermum	66	Hypoxis	82
Echium	65		
Elatine	27	Ilex	59
Elodea	27	Ilysanthes	62
Epigæa	56	Impatiens	30
Epilobium	39	Inula	50
Epiphegus	61	Iris	82
Erechthites	52	Iva	50
Erigeron	48		
Eriocaulon	85	Juncus	84
Erythronium	84	Juniperus	78

Kalmia	58	Monotropa	59
Krigia	53	Mulgedium	54
		Myrica	75
Lactuca	54	Myriophyllum	40
Lamium	65	Myosotis	65
Lappa	53		
Lapsana	53	Nabalus	54
Larix	77	Nasturtium	24
Lathyrus	33	Naumbargia	60
Lechea	26	Nemopanthes	59
Leontadon	54	Nepeta	64
Leonurus	65	Nesæa	39
Lepidium	25	Nicandra	66
Lespedeza	34	Nuphar	24
Leucanthemum	52	Nymphæa	23
Liatris	46	Nyssa	44
Ligusticum	42	Œnothera	39
Ligustrum	68	Onopordon	53
Lilium	83	Opuntia	40
Limnanthemum	67	Ornithogalum	83
Linaria	61	Osmorrhiza	43
Linnea	44	Ostrya	75
Linum	30	Oxalis	31
Liparis	80		
Lobelia	55	Panax	43
Lonicera	44	Parnassia	27
Ludwigia	39	Pastinaca	42
Lupinus	33	Pedicularis	63
Lychnis	28	Peltandra	78
Lycopus	64	Penthorum	40
Lysimachia	60	Phryma	63
Lythrum	38	Phytolacca	70
		Pilea	73
Magnolia	23	Pinus	77
Malva	30	Plantago	59
Maruta	51	Platanthera	81
Medeola	83	Platanus	74
Medicago	35	Pluchea	50
Melampyrum	63	Pogonia	81
Melilotus	35	Polygala	33
Mentha	63	Polygonatum	83
Menyanthes	67	Polygonum	70
Mikania	47	Pontederia	84
Mimulus	62	Populus	77
Mitchella	46	Portulaca	29
Mollugo	29	Potamogeton	79
Monarda	64	Potentilla	36
Moneses	58	Prinos	59

Proserpinaca	40	Sclidago	49
Prunella	64	Sonchus	54
Prunus	35	Sparganium	79
Pycnanthemum	64	Specularia	55
Pyrola	58	Spergula	29
Pyrus	38	Spergularia	29
		Spiræa	36
Quercus	74	Spiranthes	81
Raphanus	25	Stachys	65
Ranunculus	22	Statice	59
Rhexia	38	Stellaria	28
Rhodora	57	Suæda	69
Rhus	31	Symplocarpus	79
Ribes	41		
Robinia	34	Tanacetum	52
Rosa	37	Taraxacum	54
Rubus	37	Teucrium	65
Rudbeckia	50	Thalictrum	21
Rumex	71	Thaspium	42
		Thymus	64
Sagina	29	Tiarella	41
Sagittaria	80	Tilea	30
Salicornia	69	Trichostema	65
Salix	76	Trientalis	60
Salsola	69	Trifolium	34
Sambucus	45	Triglochin	80
Sanguinaria	24	Trillium	82
Sanguisorba	44	Triosteum	45
Sanicula	42	Turritis	44
Saponaria	28	Typha	79
Sarracenia	24		
Sassafras	71	Ulmus	73
Saxifraga	41	Urtica	73
Scleranthus	29	Utricularia	60
Scrophularia	61	Uvularia	84
Scutellaria	64		
Sedum	40	Vaccinium	56
Sempervivum	40	Veratrum	84
Senecio	53	Verbascum	61
Sericocarpus	47	Verbena	63
Silene	28	Veronica	62
Sinapis	25	Viburnum	45
Sisymbrium	25	Viola	26
Sisyrinchium	82	Vitis	31
Sium	42	Xanthium	50
Smilacina	83	Xyris	85
Smilax	82		
Solanum	67	Zostera	79

INDEX OF COMMON NAMES.

Aaron's Rod 40
Agrimony 36
Alder, Black 59
    Common 76
    Speckled 76
    White 57
Amaranth 70
    Mongrel 70
American Valerian 82
Andromeda, Privet 57
    Rosemary 00
    Rusty-leaved 57
Anemone, Common . . . 21
    Rue 21
    Tall 21
Angelica, Lesser 42
Apple of Peru 66
Arethusa 81
    Adder's Tongue . . . 81
Arrow-grass, Marsh . . 80
Arrow-head 80
Arrow-wood 45
    Maple-leaved . . . 45
Artichoke, Jerusalem 51
Arum, Arrow 78
    Water 78
Ash, Black 68
    Mountain 38
    White 68
Aspen, American 77
Aster, Annual Salt Marsh . 48
    Bushy 47
    Corymbed 47
    Heart-leaved 47
    Mean 48

Aster, Narrow-leaved . . . 47
    New England 48
    Pointed-leaved 48
    Rough-stemmed . . . 48
    Spreading 47
    Smooth Blue 47
    Variable 47
    White-topped 47
    Willow-leaved 48
Avens, Purple 36
    Tall Yellow 36
    White 36

Balmony 61
Barberry 23
Basil, Wild 64
Basswood 30
Bayberry 75
Beach Pea 33
    Plum 35
Bean, Wild 71
Bedstraw, Sweet 45
    Small 45
Beech 75
Beech Drops 61
    False 59
Beggar's Lice 66
Beggar-ticks 51
    Swamp 51
Bell-flower, Slender . . . 55
Bellwort, Clasping 84
    Common 84
Bergamot, Wild 64
Bindweed 66
    Black 71

Bindweed, Hedge	66	Burr Reed, Smaller	79
Birch, Black	76	Bush Clover	34
Canoe	75	Hairy	34
Cherry	76	Butter and Eggs	61
Paper	75	Buttercups	22
Red	76	Buttonbush	46
River	76	Buttonwood	74
White	75		
Yellow	76	Cancer-root	61
Birthroot	83	One-flowered	61
Bitter-sweet	32	Caraway	43
Black-grass	84	Cardinal Flower	55
Blackberry, High	37	Carpet Weed	29
Low	37	Carrion Flower	82
Swamp	37	Carrot	42
Black Larch	77	Catchfly	28
Bladder Campion	28	Catnip	64
Bladderwort, Common	61	Cat's Paw	52
Creeping	61	Cat-tail	79
Horned	61	Cedar, Red	78
Inflated	60	White	78
Purple	60	Celandine	24
Blazing Star	46	Charlock	25
Bloodroot	24	Checkerberry	57
Blueberry, Black	56	Cheesevine	30
High Bush	56	Cherry, Black	36
Low	56	Choke	35
Blue Curls	65	Wild Red	35
Blue-eyed Grass	82	Chestnut	75
Blue Flag, Large	82	Chickweed	28
Slender	82	Field	29
Bluets	46	Mouse-ear	28
Blue Weed	65	Chicory	53
Boneset, Broad-leaved	46	Chokeberry	38
Smooth	46	Cinquefoil, Crowded	36
Verbena-leaved	46	Mountain	37
Bouncing Bet	28	Norway	36
Boxberry	46	Shrubby	37
Buckbean	67	Silvery	36
Buckwheat	71	Clearweed	73
Climbing False	71	Cleavers, Rough	45
Bullbrier	82	Clintonia, Northern	83
Bulrush	84	Clover, Hop	35
Burnet, Canada	44	Low Hop	35
Burdock	53	Pussy	34
Burdock, Sea	50	Rabbit-foot	34
Burr Marigold, Nodding	51	Red	34
Burr Reed	79	Sweet	35

Clover, White	34	Dandelion	54
Wooly-stemmed	34	Dwarf	53
Zig Zag	34	False	54
Cockle-burr	50	Horse	54
Cohosh, White	23	Dangleberry	56
Columbine	23	Desmodium	34
Cone-flower	51	Diplopappus, Cornel-leaved	49
Coral-root, Large	80	Large	49
Late	80	Violet	49
Corn Cockle	28	Dock, Bitter	71
Cornel, Alternate-leaved	44	Curled	71
Dwarf	44	White	71
Panicled	43	Willow-leaved	71
Red Osier	43	Yellow	71
Round-leaved	43	Dodder, Common	66
Corn Spurry	29	Flax	66
Corydalis, Pale	24	Dogberry	38
Cowberry	56	Dogsbane	67
Cow-lily	24	Dogwood	31
Cow-parsnip	42	Flowering	43
Cowslip	22	Dutch Myrtle	75
Cow Wheat	63	Dyers' Weed	35
Crackers	28		
Cranberry	56	Eel-grass	79
Mountain	56	Elder	45
Cranesbill	30	Poison	31
Carolina	30	Red-berried	45
Cranzia	42	Elecampane	50
Creeper	32	Elm, American	73
Cress, Bitter	25	Epilobium	39
Marsh	25	Evening Primrose	39
True Water	24	Dwarf	39
Winter	25	Everlasting, Pearly	52
Crowfoot, Bulbous	22	Plantain-leaved	52
Celery-leaved	22	Sweet-scented	52
Creeping	22	Winged	52
Cursed	22		
Early	22	False Spikenard	83
Meadow	22	Featherfoil, Inflated	60
Seaside	22	Fern, Meadow	75
Small-flowered	22	Sweet	75
Tall	22	Fever-bush	72
Water	22	Feverwort	45
Yellow Water	22	Figwort	61
Cuckold-weed	51	Fireweed	52
Cucumber-root	83	Five-finger	36
Cudweed, Low	52	Flag, Large Blue	82
Cymbidium	81	Slender Blue	82

Flag, Sweet	79	Grass-of Parnassus	27
Fleabane	48	Greenbrier	82
Daisy	48	Ground Ivy	64
Narrow-leaved	49	Groundnut	33
Purple	48	Groundsel, Common	53
Salt Marsh	50		
Floating Heart	67	Hackberry	73
Fly-trap	67	Hacmatac	77
Forefather's Cup	24	Hardhack	36
Forget-me-not	65	Harebell	55
Frostweed	26	Hawkweed, Canada	54
Fumitory, Common	24	Rough	54
		Hazelnut	75
Garget	70	Beaked	75
Garlic, Wild Meadow	83	Hearts' Ease	70
Gentian, Closed	67	Hedge Hyssop	62
Fringed	67	Golden	62
Gerardia, Bushy	63	Hedge Nettle, Marsh	65
Downy	63	Hellebore, White	84
Purple	62	Hemlock	77
Seaside	62	Bulb-bearing Water	42
Slender	63	Hemp	73
Germander	65	Indian	46
Gill-over-ground	64	Hempweed, Climbing	47
Ginseng, Dwarf	43	Henbane, Black	66
Golden-rod, Blue-stemmed	49	Henbit	65
Bushy	50	Hepatica, Blue	21
Common 3-ribbed	50	Herb Robert	30
Gray	50	Hickory	74
Late 3-ribbed	50	High Water Shrub	50
Many-flowered	49	Hoarhound, Water	64
Rough or Tall	50	Hog Peanut	33
Seaside	49	Holly, Mountain	59
Slender	49	Honeysuckle	34
Smooth	49	Bush	45
Sweet	50	Trumpet	44
White-rayed	49	White	57
Willow-leaved	49	Hop Hornbeam	75
Goldthread	23	Hornbeam, American	75
Gooseberry, Short-stalked	41	Horse Radish	25
Goosefoot, Salt	69	Horse Savin	78
Maple-leaved	69	Horse Weed	48
Oak-leaved	69	Houseleek	40
Triangular-leaved	69	Huckleberry	56
Goosetongue	52	Blue	56
Grape, Common Wild	31	Hudsonia	26
Frost	32	Huntsman's Cup	24
Summer	32		

Indian Pipe	59
Tobacco	55
Turnip	78
Inkberry	59
Innocence	46
Ivy, Ground	64
Poison	31
Jamestown-Weed	66
Jersey Tea	32
Jewel-weed	30
Jointweed	71
June-berry	38
Juniper	78
Knapweed	53
Knawel	29
Knotgrass	70
Slender	71
Ladies' Slipper, Low	82
Thumb	70
Tresses	81
Lambkill	58
Larch, Black	77
Laurel, Mountain	58
Pale	58
Sheep	58
Lepidium	25
Lettuce, Canker	58
Tall White	54
Wild	54
Leverwood	75
Licorice	46
Lily, Red	83
Yellow	83
Linden, American	30
Linnea	44
Live-forever	40
Lobelia	55
Pale Spiked	55
Water	55
Locust, Common	34
Loosestrife, Four-leaved	60
Lance-leaved	60
Low	38
Swamp	39
Tufted	60
Loosestrife, Upright	60
Lopseed	63
Lousewort	63
Yellow	63
Lovage, Scotch	42
Lupine	33
Mallows, Low	30
Maple, Striped	32
Sugar	32
Swamp	32
Marigold, Water	51
Marsh Rosemary	59
Vetchling	33
Mayblob	22
Mayflower	56
Mayweed	51
Meadow Beauty	38
Fern	75
Pea	33
Rue, Early	21
Rue, Large	22
Meadow-sweet	36
Melilot, White	35
Mercury	31
Three-seeded	73
Mermaidweed	40
Mezereum	72
Milkweed, Common	68
Four-leaved	68
Poke	68
Swamp	68
Whorled	68
Milkwort, Cross-leaved	33
Double-fruited	33
Red	33
Whorled	33
Mint, Meadow	64
Mountain	64
Mitrewort, False	41
Monkey-flower	62
Moss, Golden	40
Motherwort	65
Mountain Ash	38
Mountain Holly	59
Mouse-ear	52
Mousemead	41
Mugwort	52

Mugwort, Shore	52	Parsnip, Water	42
Mulgedium	54	Partridge-berry	57
Mullein	61	Pea, Beach	33
Musquash Root	42	Pear, Prickly	40
Mustard, Black	25	Pearlwort	29
Hedge	25	Pennyroyal	64
Myrtle, Dutch	75	False	65
		Pennywort	41
Neckweed	62	Round-leaved	42
Nettle, Common	73	Pepperidge	44
Common Hemp	65	Peppermint	64
Dead	65	Persicaria, Tall	70
False	73	Water	70
Small	73	Pettimorril	43
Stingless	73	Pickerel-weed	84
Nightshade, Common	67	Pignut	74
Enchanter's	39	Pigweed	69
Small Enchanter's	39	Pimpernel	60
Woody	67	False	62
Nipplewort	53	Pine, Pitch	77
Noble Liverwort	21	White	77
Nodding Trillium	82	Pine-sap	59
Nonesuch	35	Pineweed	27
		Pink, Deptford	28
Oak, Black	74	Swamp	57
Chinquapin	74	Wild	28
Poison	31	Wooly	28
Quercitron	74	Pinweed, Large	26
Scarlet	74	Small	26
Scrub	74	Pipewort, Small	85
Swamp Chestnut	74	Plantain	59
Swamp White	74	Narrow	59
White	74	Pale Rattlesnake	81
Orache, Spreading	69	Rattlesnake	82
Orchis, Large Purple Fringed	81	Robin's	48
Pale Green	80	Seaside	59
Ragged	81	Water	80
Round-leaved	81	Poke	70
Small Purple Fringed	81	Indian	84
White Fringed	81	Pond Lily	23
Yellowish	81	Pondweed, Clasping	80
Ox-eye Daisy	52	Floating	79
		Narrow-leaved	80
Painted Cup	63	Short-spiked	79
Parsley, Fool's	42	Slender	79
Parsnip	42	Thin-leaved	79
Cow	42	Poplar, Tooth-leaved	77
Meadow	42	Prim	68

Prince's Pine	58
Privet	68
Purslane	29
Milk	73
Water	39
Pyrola, Broad-leaved	58
One-flowered	58
One-sided	58
Round-leaved	58
Small	58
Thin-leaved	58
Queen Anne	73
Queen-of-the-Meadow	46
Ragwort, Golden	53
Raspberry, Red	37
Flowering	37
Rattlesnake Plaintain	82
Root	54
Weed	54
Red Thorn	38
Rhodora	57
Rock-rose	26
Roman Wormwood	50
Rose, Low Wild	37
Rudbeckia, Cut-leaved	50
Rush, White-seeded	84
Salt Goosefoot	69
Saltwort	69
Samphire	69
Sandwort	29
Sea	28
Side-flowering	28
Thyme-leaved	28
Sanicle	42
Sarsaparilla, Bristly	43
Wild	43
Sassafras	71
Saxifrage, Early	41
Golden	41
Swamp	41
Scabish	39
Scoke	70
Scorpion-grass	65
Scratch-grass	71
Larger	71
Scullcap	64
Side-flowering	56
Sea Rocket	25
Seed-box	39
Self-heal	64
Senecio, Golden	53
Shad-bush	38
Shepherd's Purse	25
Side-saddle Flower	24
Silver-weed	39
Skunk Cabbage	79
Smartweed	70
Snake-head	61
Snapdragon, Canada	61
Sneezewort	52
Soapwort	28
Solomon's Seal, Small	83
Star-flowered	83
Two-leaved	83
Sorrel, Common	71
Ladies'	31
Yellow Wood	31
Sow Thistle, Common	54
Prickly	54
Sparganium	79
Spearmint	63
Spearwort, Creeping	22
Specularia, Clasping	55
Speedwell, Common	62
Corn	62
Marsh	62
Thyme-leaved	62
Water	62
Spice Wood	72
Spikenard	43
False	83
Spiranthes, Naked	81
Spreading Orache	69
Spurge, Leafy	73
Seaside	73
Spotted	73
Sun	72
Staff Tree	32
Star-flower	60
Star-of-Bethlehem, White	83
Yellow	82
Starwort, Narrow-leaved	72
Water	72

Stickseed	66
Stitchwort, Northern	28
St. John's-wort	27
Canadian	27
Marsh	27
Small	27
Stonecrop, Ditch	40
Mossy	40
Strawberry	37
Long-fruited	37
Succory, Wild	53
Sumach, Dwarf	31
Poison	31
Smooth	31
Staghorn	31
Sundew, Long-leaved	27
Round-leaved	27
Sunflower, Cross-Leaved	51
Pale-leaved	51
Tickseed	51
Sweet Bay	23
Brier	38
Cicely, Hairy	43
Fern	75
Flag	79
Gale	75
Pepperbush	57
William	28
Tansy	52
Tear-thumb	71
Thimbleberry	37
Thistle, Canada	53
Common	53
Cotton	53
Pasture	53
Thistle, Two-colored	53
Thorn-Apple	66
Thorn, Red	38
Thoroughwort	46
Thyme, Garden	64
Tick Trefoil	34
Tick, Naked-flowered	34
Toad-flax	61
False	72
Touch-me-not	30
Tower Mustard, Smooth	44
Trailing Arbutus	56
Traveller's Joy	21
Trumpet Honeysuckle	44
Weed	46
Tupelo	44
Turnip, Indian	78
Wild	25
Twayblade	80
Twin-flower	44
Valerian, American	82
Velvet-leaf	29
Vervain, Blue	63
Nettle-leaved	63
Viburnam, Sweet	45
Violet, Arrow-leaved	26
Bird-foot	26
Dog-tooth	84
Downy yellow	26
Hooded	26
Horse	26
Lance-leaved	26
Spreading	26
Sweet, White	26
Virgin's Bower	21
Water Milfoil, Variable	40
Lily, Yellow	24
Pepper, Mild	70
Shield	23
Water-wort	27
Wax-work	32
White Alder	57
Whiteblow	27
Whiteweed	52
Whitlow-grass, Common	44
Wild Bean	71
Cotton	29
Flax	30
Geranium	30
Indigo	35
Peppergrass	25
Pink	28
Radish	25
Turnip	25
Willow-herb, Great	39
Purple-veined	39
Willow, Hoary	76
Low Bush	76

Willow, Pussy	76	Woodbine	32
Swamp	76	Wood Sage	65
Yellow	77	Wood Waxen	35
Wintergreen, Spotted	58	Wormseed	69
Witch Hazel	41		
Withe-rod	45	Yarrow	52
Wood Betony	63	Yellow-eyed Grass	85

www.ingramcontent.com/pod-product-compliance
Lightning Source LLC
Chambersburg PA
CBHW020158170426
43199CB00010B/1091